SCIENTIFIC AMERICAN

Inventions

from Outer Space

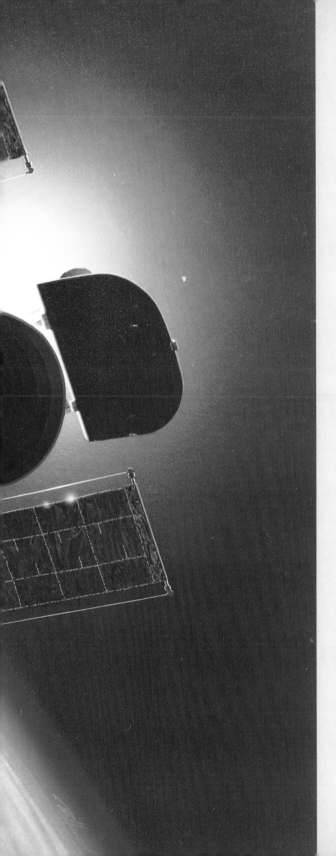

SCIENTIFIC
AMERICAN

Inventions

from Outer Space

Everyday uses for
NASA technology

David Baker

RANDOM HOUSE
NEW YORK

Scientific American
Inventions from Outer Space: Everyday Uses for NASA Technology

Copyright © 2000 Marshall Editions Developments Ltd.
Copyright © 2000 Random House, Inc.

Produced in association with Random House Reference & Information Publishing

Project Editor Conor Kilgannon
Managing Editor Clare Currie
Managing Art Editor Patrick Carpenter
Editorial Director Ellen Dupont
Production Amanda Mackie
Editorial Coordinator Ros Highstead

Produced for Marshall Editions by **Design Revolution**
Queens Park Villa, 30 West Drive, Brighton BN2 2GE
Editors Ian Kearey, Ian Whitelaw
Designer Lucie Penn
Art Editor Andy Ashdown
Picture Research Penni Bickle
Indexer Indexing Specialists, Hove, East Sussex

Originated by Master Image, Singapore
Printed and bound in Italy by Milanostampa

Library of Congress Cataloging-in-Publication Data
Baker, David, 1944–
  Scientific American inventions from outer space: everday uses for NASA
technology/David Baker.
     p. cm.
    ISBN 0-375-40979-3
    1. Inventions 2. United States National Aeronautics and Space Administration. 1.
Title
T212 B33 2000
609--dc21

                                                                99-055021

First Edition
0 9 8 7 6 5 4 3 2 1
February 2000
SAP Network: 10038328

New York Toronto London Sydney Auckland

CONTENTS

NASA
INVENTIONS

On october 1, 1998 nasa celebrated 40 years of extraordinary human achievement in the peaceful exploration of space.

Since 1958, NASA has carried the first human beings to the Moon, explored eight of the nine planets in our solar system, and set up the world's first reusable space transportation system. Now, at the start of the new millenium, it has begun to assemble the international space station—an orbiting complex the size of a football field involving more nations working together as a team than any other project in history. The space station will undoubtedly focus minds on the needs of our planet, but NASA is not a newcomer to improving the lives of people on the Earth, as the following pages testify. As NASA looks forward to the future, it is building on the specific focus of each of the four preceding decades.

1960s EXPLORATION

1970s APPLICATION

1980s OBSERVATION

1990s UTILIZATION

From unmanned robots exploring the moon and the planets to human bootprints on the lunar surface, NASA forged a place in history during this first decade of the Space Age. Giant rockets, powerful satellites and intelligent robots all came of age to create a new Space Age, promising innovative and exciting products for everyday life. It was from these individual achievements that a rich harvest of space products for Earth-based use came into being. NASA also decided to build a reusable Shuttle for more efficient space transportation.

NASA began to develop satellites for Earth observation, for monitoring the planet's resources, and for the detection of pollution. After discovering the ozone layer that protects life on the Earth from harmful ultraviolet rays, satellites detected the pollutants that were eating this important layer away. After measuring the effects of overheating on the surface of Venus, scientists were able to unravel the effects of toxic chemicals on the Earth's environment. By the end of the decade, satellites designed to expand knowledge of our planet had given a new awareness of our vulnerability.

The Shuttle was used to install new instruments on this orbiting telescope and give Hubble a better view of space. The Great Observatories were launched, including the Hubble Space Telescope, which provided new evidence about the origin of the universe. NASA introduced the Shuttle and built up flights to 10 a year, before losing seven astronauts when Challenger exploded, delaying the program by nearly three years. Throughout the decade, NASA expanded its observatories program and extended its observation of the Earth's resources.

NASA mobilized a group of nations, including the former Soviet Union in the assembly of the international space station, the biggest cooperative science and engineering project of all time. Working to a new mandate of "better, cheaper, faster," NASA switched its efforts from large projects and giant spacecraft to more focused projects and smaller spacecraft launched more frequently. Becoming a leaner, less costly, and more efficient organization, NASA set the stage for major new discoveries aboard the space station during the first decade of the new century.

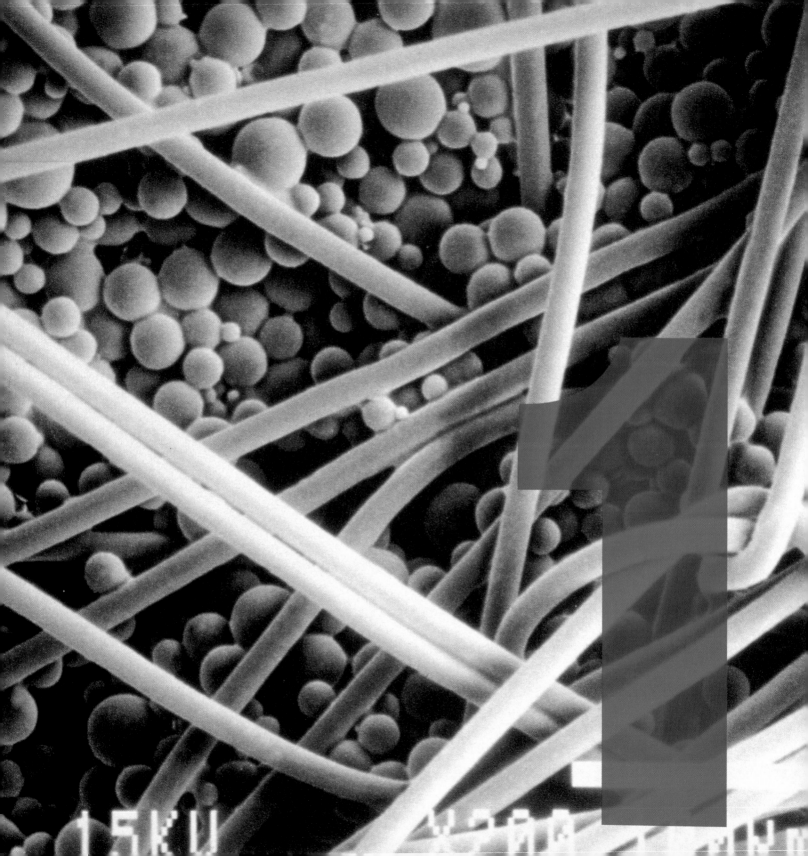

HEALTH, MEDICINE, AND PUBLIC SAFETY

HEALTH AND SAFETY HAVE BEEN VITAL PARTS OF NASA'S PROGRAM SINCE IT FIRST TOOK CONTROL OF THE U.S. GOVERNMENT'S NONMILITARY SPACE PROGRAMS IN 1958. THESE CONCERNS HAVE SPILLED OVER INTO APPLICATIONS THAT DIRECTLY BENEFIT PEOPLE ON EARTH.

Many health applications originate in the imaginations of engineers, technicians, or scientists, and are developed by commercial health-care organizations. The heart-imaging system (see p. 14) is a direct spin-off from work already under way to understand the way the human body reacts to weightlessness. Other applications, such as the physical fitness machine (see p. 18), stem from microtechnology developed for a piece of hardware designed to fly in space. There is even a system to alert rescue centers when adventurers get into trouble far from home (see p. 32).

Overexposure to harmful solar rays can seriously damage eyesight, often leading to a condition called "senile macular degeneration" of the eye which impairs the vision of elderly people. A sunglasses company now believes it can make a real contribution to the worldwide effort to improve eyesight in the old and the ill by providing protection from the most damaging rays.

NASA FILTERS OUT HARMFUL RADIATION TO STUDY THE SUN

NO RADIATION

Most sunglasses sold for comfort and eye protection are not up to the job of shielding the eyes from all harmful electromagnetic radiation, but true radiation blockers like these filter more than 99 percent of damaging solar rays.

RADIATION BLOCKERS FOR SUNGLASSES

Sometimes the multiple applications of space research just keep on coming. The Suntiger series of sunglasses, manufactured by the Biomedical Optics Company of America, Inc., show how one NASA spin-off leads to another. Located in North Hollywood, California, BOCA acquired the rights to sunglasses designed to enhance eye protection and reduce the risk of cataracts and other age-related vision impairments.

These sunglasses trace their roots to NASA research from the early 1970s on advanced optical coatings for mirrors and lenses for cameras and telescopes used in space. Scientists working at the Jet Propulsion Laboratories of the California Institute of Technology recognized the industrial applications of their research. James B. Stephens, an engineer, teamed up with the late Charles G. Miller, a physicist, to design a **dye curtain** to filter out harsh blue and ultraviolet light for full-time welders. Working for three years in their spare time, the two men produced an effective screen that was marketed and sold widely to the industry.

Inspired by this, fellow NASA workers Laurie Johnson, Paul Diffendaffer, and Charles Youngberg eventually produced an **advanced filter** for protecting eyes from harsh sunlight. The team used a computer to identify hazards from different light frequencies, and incorporated their acquired medical knowledge from the dye curtain project into the design of a family of sunglasses that could protect people from intense sunlight and also screen them from harmful artificial light in factories and offices.

Marketed under the brand name of Suntiger, these lenses filtered out 99 percent of all blue, violet, and ultraviolet radiation from the sun and gave a much clearer and **more distinct image** than conventional lenses. In 1991 BOCA acquired the rights to the Suntiger line. All photo-wavelengths believed to be harmful to the human eye are now screened out by BOCA's Eagle 475 line.

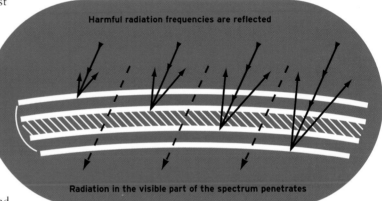

Harmful radiation frequencies are reflected

Radiation in the visible part of the spectrum penetrates

⊕ **A number of layers in the lens reflect different parts of the spectrum, effectively blocking radiation harmful to exposed eyes. Together, the combined layers work as a radiation screen while still allowing most of the visible light to reach the eyes.**

SPACE PUMP THAT
CONTROLS DIABETES

SATELLITE DEVELOPMENT LEADS TO MICROMINATURIZATION

Many diabetes sufferers must take regular doses of insulin daily to counteract the effects of a malfunctioning pancreas and maintain necessary levels of sugar. In the United States thousands have freed themselves from a rigid timetable of injections with pump therapy that uses technology developed by NASA for small satellites.

Developed in the early 1980s, pump therapy makes use of an external pump that delivers insulin at a continuous and preprogrammed rate, adjusted according to the needs of the individual. A big advantage of the pump system is that diabetics can receive a continuous flow of tiny amounts of **"short-acting" insulin** instead of getting a massive shot of "long-acting" insulin all at once. This helps the body to stabilize insulin absorption. The dosage is given at a rate that closely approximates the natural production in a healthy pancreas.

Beginning in the early 1980s MiniMed Technologies of Sylmar, California, produced an infusion pump designed to replicate as closely as possible the action of a normal pancreas. The pump was based on microengineering technology that had been used to create tiny components for environmental control in satellites and spacecraft. **No larger than a credit card**, the pump weighed only 3.7 ounces (107 g) and housed a microprocessor, a long-life battery, and a syringe reservoir filled with insulin. The syringe was connected to an infusion set: a flexible tube approximately 1 yard (1 m) long, with a needle at the end that was inserted under the skin, usually in the abdomen. Clipped to a belt or piece of clothing, the pump infused insulin at a programmed rate set by small push buttons.

BACK IN THE FAST LANE

Having to take a regular dose of insulin can interfere with the normal daily life of people with diabetes. A system that controls and delivers proportional amounts of insulin automatically on a continuous basis allows the diabetic to enjoy life to the fullest. *Inset*: An external MiniMed pump, attached at the waist, delivers insulin at a preset rate through a flexible tube.

As microengineering advanced, NASA created a new generation of planetary exploration vehicles. Based on the technology used for soil sampling and analysis equipment, MiniMed developed an implantable insulin pump that frees patients from any outside supply. Powered by a battery that lasts three years, the pump contains a sufficient quantity of insulin to last three months. The MiniMed 2001 automatically delivers insulin at a personally calibrated rate, programmed electronically by radio signals outside the body that are encrypted to ignore any other command. The insulin supply can be refilled by inserting a needle through the skin into a fill port of the **negative pressure reservoir.** When the needle is properly seated, the vacuum inside the pump automatically draws insulin from the syringe.

YIELDING FOAM ABSORBS G-FORCES OF ASTRONAUTS AT TAKE-OFF

A comfortable bed is not something you would immediately associate with the work of NASA. However, a range of revolutionary mattresses that alleviate various health problems owes its success to a material developed to help relieve the effects of G-force on astronauts.

In the late 1950s, NASA scientists began to develop a material that could be used inside spacecraft to relieve the **enormous G-force** experienced by astronauts during lift-off and flight. The result was a new kind of foam that yielded to the body and distributed its pressure evenly around the area of body in contact with it. The new material was used in both space research and the aeronautical industry, but it was not until the 1980s that the benefits of this futuristic foam found ground-based application.

Swedish company Fagerdala World Foams was the first commercial organization to realize the potential of NASA's anti-G-force foam. Fagerdala spent ten years developing

RESTFUL SLEEP

A clinical study by The Institution for Clinical and Physiological Research, Lillhagen, Sweden, found that subjects turned an average of 17 times a night on the Tempur bed but between 80 and 100 times on ordinary mattresses.

G-FORCE PROTECTION RELIEVES BACK PAIN

the material and, after 5,000 registered trials, succeeded in attaining a high-capacity production process for the manufacture of its product. The new commercially available foam was called Tempur Material and was produced by Tempur-Pedic Inc.—one of several companies that come under the Fagerdala World Foams umbrella.

Tempur Material has been heralded as the **biggest breakthrough in sleep technology** for more than 70 years. Orthopedic research has shown that, contrary to conventional wisdom, mattresses should be supportive rather than hard, because during the night we adjust our sleeping position as pressures and tensions within the body are relieved. Since the 1920s, mattresses have been internally sprung and, all too often, these conventional beds contort the sleeper into unnatural positions as he or she adjusts to the mattress. The result is a night of tossing and turning to try and get comfortable. However, the combination of NASA technology and Swedish ergonomic ingenuity has led to a yielding bed which conforms to the shape of the individual. Tempur mattresses support the body in its natural anatomical position without being too firm or too soft. The success of this NASA-inspired sleep system stems from the shape of the cells that form the structure of Tempur Material. Conventional polyurethane mattresses have irregularly shaped cells that compress under weight. However, Tempur Material has spherical cells that react to body heat and weight, and which, when displaced, reorganize their position to conform with the contours of the sleeper's body, thereby cushioning joints and giving maximum pressure relief. Tempur mattresses are also self-ventilating and allow air to move freely through them.

The unique pressure-relieving properties of Tempur Material enable it to help alleviate the symptoms of a variety of common ailments. In particular, it has been used effectively to relieve back, neck and joint pains, which are often caused or aggravated by unnatural sleeping positions. Tempur mattresses have also been **successful in alleviating pain** caused by medical conditions such as sciatica, fibrositis and circulatory problems. Such has been the success of this new material that thousands of doctors, chiropractors, physiotherapists and osteopaths now recommend that their patients use Tempur mattresses.

A yielding Tempur Material pillow helps prevent unhealthy curving of the spine at night by shaping itself to the head and neck.

THERMAL INSULATION

If you use a boiler that is so cool on the outside you would never know it was on, you may be using next-generation Space Shuttle technology. Insulation developed to cope with extremes of heat and cold in space is now finding many other uses on earth.

KEEPING IN THE HEAT

This highly efficient thermal glove offers comfort and safety at even the lowest temperatures. It incorporates a flexible insulation material combining metal alloys and ceramics, initially developed by NASA to protect shuttle craft of the future against the cold of space and the fierce heat of reentry.

When the Shuttle returns to the Earth it has to withstand temperatures of almost 2,900°F (1,600°C), generated through friction as it reenters the atmosphere at 17,520 miles (28,200 km) per hour. Protected by the insulating effect of **thermal protection tiles**, the vulnerable aluminum structure and the human occupants it carries are not affected by these temperatures. And this protection can be used many times—unlike the materials used on earlier manned spacecraft missions, which could be used only once. While advanced heat-sink materials such as the ceramic tiles used on the Shuttle absorb heat and dissipate energy gradually, without the tile burning away, they have the disadvantage of being brittle and easily damaged.

To counter this, a more robust heat shield for future spacecraft has been designed and put into production. It is composed of an improved, lightweight insulation fabricated from special **metal alloys and ceramics**. It restricts both convective and radiative heat transfer and can be made into a solid structure or woven into a blanket. Its lightness is important because it reduces the amount of weight that has to be carried into space; lower weight means smaller rockets and lower cost.

Through a company in Arizona that received a small business contract from NASA to develop this lightweight insulation for spacecraft, commercial products for sports, leisure, and paramedic uses have evolved. Thermalon Industries Ltd., in El Segundo, California, is marketing the insulation through a wide range of products. Because it retains heat four times as effectively as wool, as well as being nonallergenic and drying five times faster, it is used as a **lightweight plastic insulation** for blankets and mittens. Rescue services have found it useful as an insulator for emergency blankets, and Thermalon expects to distribute more than 70,000 of these annually; ambulance companies and the Red Cross have singled the material out as particularly useful. The mittens are designed to protect people's hands in extreme conditions, and they have already been used for industrial, military, and recreational purposes.

The company is currently working in association with NASA's Ames Research Center to find improvements that will broaden the insulation's uses for emergency rescue and care. Other spin-offs include use as a **thermal insulator** in refrigerators, replacing harmful, environmentally toxic chemicals; and for industrial and domestic boilers that experience extremes of temperature and thermal energy as part of everyday working. Aircraft also need insulation in a wide variety of forms. Because it is flexible and can be shaped easily, the new material is ideal for the areas of a jet engine that become extremely hot and need to be insulated from adjacent parts.

⬆ **Revealed by an electron microscope, the fibrous web of a composite thermal-insulation blanket can be clearly seen. It has many applications for industry, the home, and leisure.**

FURTHER USES

- ○ Heat-insulating clothing
- ○ Fire protection equipment
- ○ Water immersion survival gear in subzero conditions
- ○ Protection from toxic chemicals
- ○ Fire guards for use in the home and in the workplace

WORKOUTS FOR SPINAL INJURY VICTIMS

MICROCIRCUITRY FOR THE SHUTTLE ROBOT ARM SIMULATOR

People spend more on health and fitness machines than is spent on space projects. Yet the space program has helped make physical fitness possible for those crippled by spinal injuries. NASA developed a neuromuscular electrical stimulation device that is used regularly by many sufferers, including *Superman* star Christopher Reeve, who was paralyzed from the neck down after falling from a horse.

The story of the **muscle stimulant** begins in the 1970s, when NASA was developing the reusable Shuttle. Designed to carry large payloads into space and put together a permanently habitable space station, the Shuttle needed a robot arm to grapple payloads and modules in and out of the cargo bay. Called the Remote Manipulator System, or RMS, the enormous arm was built by Spar Aerospace in Canada. The length of two telegraph poles placed end to end and weighing 992 lb (450 kg), the RMS is capable of handling weights of up to 30 tons (30,480 kg), maneuvering them in and out of the cargo bay so gently that in operating tests astronauts were able to bring the arm to rest on top of an egg without breaking the shell.

To properly train astronauts to use the RMS, NASA built a special simulator. Weighing 2.16 tons (2,200 kg), this device accurately reproduces all the operating modes of a Shuttle arm and contains **microminiaturized circuitry** to make it accomplish a wide range of simulated tasks. Electrologic of America (ELA), based in Dayton, Ohio, took the densely packed electronics engineered for the simulator and applied them to electrical stimulation devices for medical markets, coming up with the Stim Master Ergometer exercise machine. This device contains gears hooked up to an **adaptive feedback system** that senses the pressure and muscular resistance from the operator 40 times a second and continually feeds it to a computer, which then adjusts tension and resistance.

For paraplegics and quadriplegics, one session with Stim Master provides a full cardiovascular workout equal to jogging three miles three times week. Users have found that it helps to improve circulation, relax muscle spasms, and stimulate leg muscles. The machine is designed for home as well as clinical use. Taking the microelectronic circuitry and the Stim Master technology a step further, ELA have produced a portable electrical stimulation device called the VST-100. The size of a small suitcase, it delivers tiny electrical signals that can, for instance, increase circulation in a damaged wrist by **opening nerve pathways** and speeding a return to normal.

MOVING MUSCLES

Even when disabled through spine or nerve injury, the body still needs a daily workout to preserve a healthy heart and limbs. Using this adaptive feedback device, paraplegics can now safely exercise muscles that might otherwise waste away through lack of stimulation.

IMAGE ENHANCEMENT MAKES THE MOST OF SATELLITE DATA

Regular eye screening helps detect the kinds of abnormalities that frequently lead to a deterioration in vision. Testing is one of the simplest things to do, yet it can actually save some young children from going blind. There was little chance of conducting these tests on a wide scale, however, until a NASA device made them possible.

OCULAR SCREENING SYSTEM

There are thought to be more than 15 million blind people in the world today. Less than 15 percent of that total are children, while more than half are over 60 years old. The progressive onset of **poor eyesight and blindness** in half of those who suffer serious visual impairment could be eliminated with routine checkups to measure reflective qualities of the retina at an early age. The technology from which NASA obtained crystal-clear pictures of the Moon has been adopted by Vision Research Corporation of Birmingham, Alabama, to produce a quick, inexpensive, and trouble-free way of vision-screening for the very young—and it can help older people, too.

In the mid-1960s NASA's unmanned Ranger probes were sent on a collision path with the Moon. Just before impact, cameras took pictures that scientists put through special **image-processing techniques** to enhance small details and make fuzzy pictures appear bright. Nothing is put into the image that is not already there, but objects that blend in with the background can be made to stand out so that they become readily observable and easy to identify. NASA's Marshall Space Flight Center subsequently developed an enhanced image-processing technique for eye tests, through an engineering applications project—a routine program at NASA centers in which space technology is "brainstormed" out of the door and into everyday life.

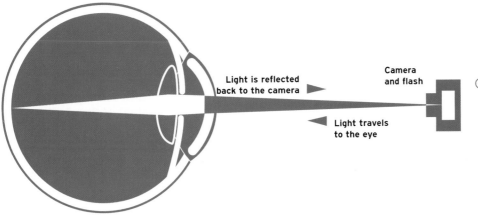

Light is reflected
back to the camera ▶

Camera
and flash

◀ Light travels
to the eye

The bright light from a
photoflash penetrates the
eye and is reflected by the
retina back to the camera,
which captures an image
that can then be analyzed.

In principle, the eye-screening technique from Vision Screening is simple. A standard 35mm camera with telephoto lens and electronic flash is located in the back of a box placed approximately 6 feet (2 m) away from a small square window. The person places his or her face against the window, and when the camera takes an exposure, the flash illuminates the eyes. The reflected image from the retina is captured on film, and a photorefractor analyzes an enhanced picture to produce a **color view of the retina**. Specialists examining the image can detect the early stages of amblyopia, or "lazy eye," which causes gradual dimming, and sometimes loss, of vision.

In the early 1990s about 50,000 children in the United States were screened for retinal defects using the NASA technology. About 4,000 children received **immediate treatment** that may prevent the majority of them from going blind.

MAKING EYE TESTS EASIER

It is difficult to get children to remain still for the time needed to perform a standard eye-screening test, but with this ocular screening system the child only has to stay still long enough for a picture to be taken.

SPRAY-ON FOAM KEEPS THE SHUTTLE'S FUEL SUPERCOOL

A design solution originally created to keep Shuttle propellants at supercold conditions before they are ignited is now helping to improve the manufacture and fitting of prosthetic devices. It can do this by enhancing the accuracy of the mirror-image mold of the patient's residual limb.

MATCHING FOR COMFORT

For those needing a prosthetic device, everyday life is made easier by a type of foam designed to protect the Shuttle tank from overheating. The foam, polyisocyanurate, is used to register a precise mold of the patient's stump, enabling a more accurate match to be made, thus ensuring a greater degree of comfort.

PROSTHESIS FROM THE SHUTTLE TANK

In the late 1960s NASA decided to use the highly efficient combination of liquid hydrogen and liquid oxygen propellants to power the reusable Shuttle into orbit. When ignited in a rocket engine these propellants produce more energy than almost any other combination of liquids, but to carry them in the minimum volume possible they must be stored in a liquid state, which requires supercooling. The supercold liquids must be protected from heat at all times, whether it is from the hot Florida sun as the shuttle is prepared for the launch, the increase of temperature due to friction as the craft speeds toward orbit, or the superhot gases burned in the main engine. The problem, however, is that liquid oxygen boils at 147°F (-46°C) and liquid hydrogen boils at -217°F (-102°C), so they need a specific type of very efficient insulation.

To protect these supercold liquids the two propellants are each contained in what amounts to a pair of **giant thermoses**, one stacked on top of the other in a single structure known as the External Tank (ET). This carries 385,200 US gallons (1,230,000 liters) of hydrogen and 143,300 US gallons (541,310 liters) of oxygen, and the two fluids are fed to the engines at the back of the Shuttle orbiter via separate pipes. The ET is covered with almost 2 tons of **spray-on foam insulation** (SOFI), a polyisocyanurate material that has **higher temperature stability** than conventional urethane foam. The depth of the insulation varies according to which areas need it the most. It can be as thick as 5/8 in (1.5 cm).

Working with Lockheed Martin, NASA's Marshall Space Flight Center made SOFI available to the Harshberger Prosthetic and Orthotic Center in Birmingham, Alabama. SOFI, with its high degree of temperature stability, is less subject to distortion than conventional materials and was found to be ideal for making a mold of the patient's residual limb to which a new prosthesis could be attached, providing a more exact, lighter, and more rigid mold. A foam blank is inserted into a computer-controlled fabricating machine, which then carves an exact mirror shape of the patient's stump, ensuring a snug fit that improves comfort and reduces abrasion of the residual limb.

FURTHER USES

- Precision molding for engineering components
- Accurate pattern-matching for manufacturing industries
- Better oven insulation
- Improved molding for seats

The marvelous advances of modern medicine are sometimes achieved by drawing on discoveries in other scientific areas. Michael Vannier brought two disciplines together by seeing how better diagnostic techniques could be achieved using space technology. Today the technique he pioneered is standard, helping make diagnosis more rapid and speeding treatment.

BODY SCANNING
IMPROVES DIAGNOSTICS

The story began when NASA developed the first satellite remote sensing system toward the end of the 1960s. Called Landsat, the satellite orbited the Earth at a height of 559 miles (900 km), crossing both poles, with an electronic camera sending images to receiving stations on the ground. On each successive orbit the spinning Earth shifted eastward almost 1,800 miles (2,900 km), allowing the satellite to build up a **sequence of image strips** for observing the planet on a daily basis. To get the most from these images the camera observed the Earth in several segments of the spectrum, with detectors recognizing unique "signatures" of various features—crops, water, buildings, forests, etc.—in a system called "thematic mapping."

Dr. Vannier was one of several thousand physicians using **nuclear magnetic resonance** (NMR) scanning to observe sections of the human body. NMR employs a magnetic field and radio waves to create body images from which specialists can determine the severity and physical extent of a tumor or a blood clot, for example. Unlike X-ray machines, NMR does not expose the organs to radiation; however, NMR scanners collect a lot of redundant data and complicate the picture by constructing a highly detailed but complex image, and radiologists may need up to **50 scans** to make a proper diagnosis.

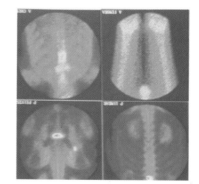

⊕ **These four NMR bone scan images on a computer screen show how thematic mapping enables physicians to distinguish quickly between different types of tissue by their color.**

Dr. Vannier, who began his career as an engineer with NASA, saw the solution in the NASA **computerized image-processing system** employed to sharpen contrasts, eliminate confusing detail, and enhance the resolution of Landsat pictures. With the help of Bob Butterfield, then manager of technology integration at NASA Kennedy Space Center, he took sample NMR scans to the Sensing and Image Processing Laboratory at the University of Florida and subjected them to the Landsat enhancement techniques. The results led to the team producing colored "theme" maps from the scans, allowing physicians to see the exact profile of the diseased tissue immediately.

FULL-BODY SCANNING

Images from this nuclear magnetic resonance body scanner are passed through filters to reduce the number of scans necessary for a positive diagnosis from as many as 50 to just one.

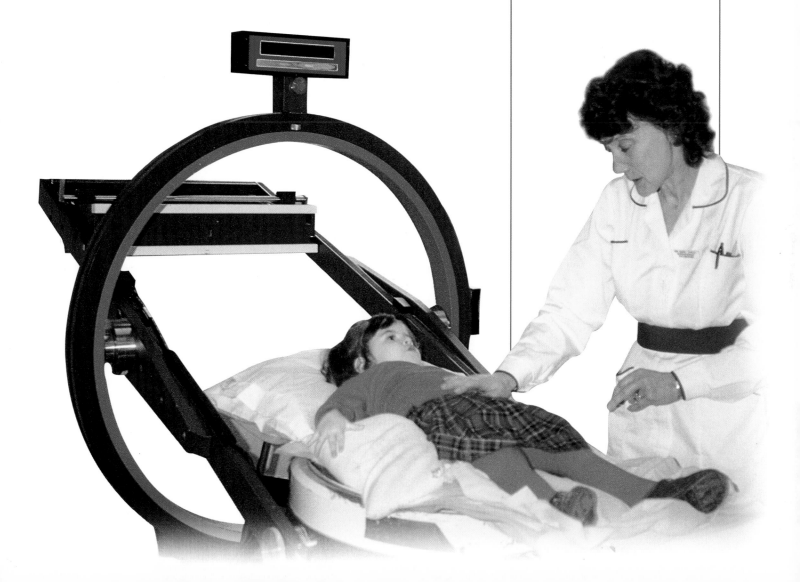

CARDIOMUSCULAR CONDITIONER

If staying fit on Earth is challenging, consider the difficulties of exercising effectively in a weightless state. Exercises developed for astronauts in space are now used by national football teams, university fitness coaches, sports clinics and medical rehabilitation centers for maximum workout efficiency.

MEDICAL CHECK UPS REVEAL THE
FULL EFFECTS OF WEIGHTLESSNESS

FROM SPACE TO EARTH

Exercise machines, such as this one, are used by astronauts in space to maintain fitness and strength while living in an environment of weightlessness. Technology developed for conditioning the human body in orbit has been applied to a cardiomuscular traction device that owes its origin to a design for space stations.

In the early 1960s, while the recently formed NASA was hard at work building Apollo for manned Moon landings later that decade, the US Air Force planned a space station. It was to be a research base for military astronauts, an observation platform, and a place from which reconnaissance and surveillance missions could be mounted. Medical tests on astronauts who had spent several days in space revealed serious physical deterioration. Bones lost calcium and became brittle, blood cells changed shape and disappeared, the heart became flabby and relaxed, and muscles lost their tone. Nobody knew how these effects would change with time spent in space because no one had spent more than a couple of weeks in orbit. To plan for the worst, the Air Force gave Boeing a contract to devise and build a cardiovascular exercise machine that astronauts could use in space to keep their heart and blood flow in condition. Boeing's Gary Graham was one of those who worked on the device.

At the end of the 1960s the Air Force orbiting laboratory project was canceled. Graham shifted his work to NASA's Skylab space station, which was launched in 1973 and occupied by three teams of three astronauts until it was finally abandoned in February 1974. The exercise device never made it to Skylab, but in 1987 Graham decided to revive the idea and design an earth-based version. He teamed up with exercise physician Gary Chase and refined an initial design, naming it the CMC Shuttle 2000. In the years since the original research, much more information on the physical adaptation of the human body to prolonged weightlessness had been gleaned from the long-duration Skylab flights. The results from these missions, which lasted up to 84 days, led to the design of a more effective cardiac-conditioning exerciser for everyday use. The refined device, which was completed by Graham and Chase in 1991, was manufactured and marketed by the Contemporary Design Company of Glacier, Washington, as the Shuttle 2000-1.

The Shuttle 2000-1 is based around a horizontal trampoline and is designed for plyometric exercises that stress the body and produce faster and stronger muscle contractions. The user lies on a cushioned carriage that glides between two rails, propeled off a kick plate. Resistance is provided by elastic cords running along the machine, and the user can add or remove the cords to increase or decrease resistance. The rebound feature combines both internal and external stresses. The acceleration load caused by the rebound system induces a movement of the diaphragm that can change the volume of the chest cavity, attracting more blood to the heart.

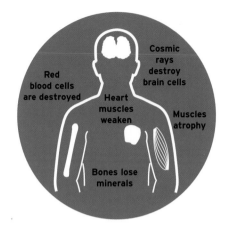

Cosmic rays destroy brain cells

Red blood cells are destroyed

Heart muscles weaken

Muscles atrophy

Bones lose minerals

Extended periods in space, exposed to cosmic radiation and conditions of weightlessness, have been found to affect many aspects of the body system, including the heart, muscles, brain, blood, and bones.

FURTHER USES

O Body toning for athletes
O Postoperative recuperation
O General health care for bedrest patients
O Measuring performance in individual organs

About one person in ten suffers from Seasonal Affective Disorder–a condition caused by lack of sunlight. Many of those affected develop clinical depression and fail to cope with everyday life. However, a light-emitting visor offers sufferers the chance to overcome this syndrome.

ASTRONAUTS SUFFER WITHOUT THE DAILY RHYTHM OF LIGHT AND DARK

FITTING A LIGHT VISOR

This NASA-inspired light-emitting visor floods the face with the right kind of light to offset the effects of dark days or long winter nights.

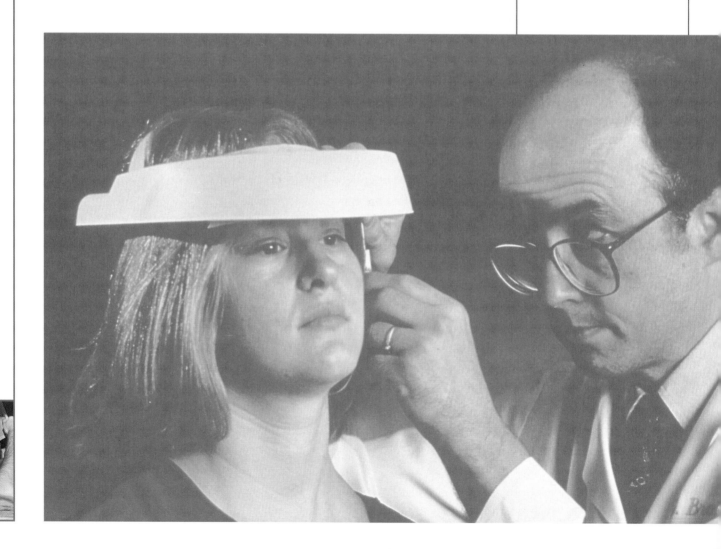

BEATING SEASONAL AFFECTIVE DISORDER

NASA became interested in the uses of light therapy when dealing with astronauts who found it difficult to maintain their normal **sleep-wake cycle** in space. When this cycle—which is also known as the circadian rhythm—is upset, body chemicals get out of balance and need to be adjusted by regulating exposure to light. Other NASA-based research revealed that people are sensitive to a certain part of the visible band of light detected by the human eye.

The circadian rhythm is set by a "clock" in the brain, which is in turn controlled by the visible band of light. The eye is, therefore, used not only for seeing but also for setting the day–night cycle for the whole body. Without light, the eyes and brain cannot regulate the body's chemical balance and set the daily clock. Researchers discovered that the effect of changes to the circadian rhythm in astronauts can be alleviated by **light therapy—**a practice that uses timed exposure to light to change the flow of certain chemicals in the body in order to alter mood and performance.

While working as a consultant with NASA, Dr. George C. Brainard, Associate Professor of Neurology at Jefferson Medical College, Philadelphia, put together a team of specialists to examine the effects of disruption to the circadian rhythm. Through this he developed a device that relieves the worst effects of Seasonal Affective Disorder (SAD)— also known as the winter blues. By shining a light across the **rim of a topless hat** called the Light Visor, some of the sunlight "lost" during the dark winter months is replaced.

Dr. Brainard and Dr. Norman E. Rosenthal of the National Institute of Mental Health have teamed up with a company called BioBrite to embark on a three-year clinical test of the Light Visor at a number of medical institutes. Dr Brainard is also conducting further research to see how the Light Visor can help alleviate the worst effects of **jetlag and working the night shift**. He hopes that other irregularities in human performance caused by interruptions to the sleep–wake cycle may also be overcome with the use of this remarkable device.

⬆ **To help offset the effects of dark winter mornings this night light comes on softly before dawn and mimics the early light from a summer sun to help counter Seasonal Affective Disorder.**

SHUTTLE FLIGHT DECK IS GUARDED
BY FIRE-DETECTION EQUIPEMENT

The modern world is full of flammable chemicals, and firefighters need to have detailed information about particular fires to do their job effectively. NASA has been active in adapting remote sensing technology to this task. For almost 20 years infrared scanning devices have been used aboard aircraft to assess the dimensions of any fire.

SEEKING OUT DANGER

Firefighters frequently use scanning devices in hazardous areas to locate the source of visible or invisible flames, or to find trapped people.

INVISIBLE FLAME DETECTOR

In January 1967 three Apollo astronauts paid with their lives when NASA miscalculated the dangers of putting people in a highly pressurized, pure-oxygen environment where flames burn brighter and move more quickly than in a sea-level atmosphere. NASA was shocked by the speed with which the fire took hold, and redesigned its spacecraft with **non-flammable materials and fireproof walls**. However, NASA still uses a wide range of dangerous materials, and operates in hazardous environments where dangerous invisible flames are a very real risk. Special care and attention is needed to minimize the risk of human catastrophe.

At the John C. Stennis Space Center engineers develop rockets for a wide variety of propulsion programs that burn hydrogen and oxygen. **One million gallons of liquid hydrogen** are used every month in the rocket motors developed at Stennis. While these liquid gases are a favored combination for efficient motors, this mixture can burn intensely with an invisible flame. To detect such fires, engineers developed FIRESCAPE, an infrared scanning device with the ability to provide sophisticated computer analysis of its readings.

Market research for the FIRESCAPE device was handled by a local company, SafetySCAN of Buffalo, New York. SafetySCAN had the foresight to imagine many applications for the scanner. With no moving parts and weighing less than 5.5lb (2.5 kg), the device is used like a pair of binoculars. The optics are sealed from contamination by smoke or grit. The scanner can be charged and running within five seconds, and can be used for two hours before needing to be recharged. It uses nothing more than a **standard lead and television monitor** as external hookups. Another great advantage is that it is well within the budget of most fire departments, especially because it has multiple uses. FIRESCAPE can be used to detect unconscious people in smoke-filled rooms and to map the very heart of a fire burning in a small space. It also can be used for assessing large-scale forest fires as well as industrial conflagrations.

⬆ **The FIRESCAPE scanner maps heat to detect different intensities that may not yet have erupted into flame.**

One of the most beneficial by-products of the space race, SARSAT/COSPAS has directly saved more lives than any other application from the space program. It is a search-and-rescue system for people in danger or distress, which uses technology originally developed for monitoring weather patterns.

SEARCH AND RESCUE FROM SPACE

SATELLITES CARRY TRANSPONDERS TO RELAY EMERGENCY CALLS

The meteorological satellites NASA developed in the 1960s allowed for a routine observation of the Earth's weather, part of an international effort to understand daily changes in the atmosphere and their relationship to climate. The Soviet Union followed quickly on the heels of NASA and built its own series of meteorological satellites. Through programs put together for **global atmosphere and climate monitoring** came international cooperation that soon embraced Japan and China, who had also developed their own weather satellites. The result was a series of agreements between the space-faring countries to pool their research and information in a common knowledge base.

France, who had previously partnered the Soviet Union in several science ventures, developed a plan with Canada to set up a **global search-and-rescue system,** which utilized satellites built and launched for the global weather watch. By using the indigenous communications equipment and a small electronic package attached to an exterior part of each satellite, signals sent from beacons carried by ships and aircraft in distress would be instantly monitored and rescue services alerted. Ships and commercial aircraft, as well as many small private planes,

⚐ **Complex satellites carry small strap-on packages tuned to listen for emergency calls.**

carry these distress beacons, but until the development of the SARSAT/COSPAS system (SARSAT stands for "search-and-rescue satellite," and COSPAS is the Russian equivalent) in 1982 there was no continuous monitoring system worldwide and therefore no guarantee that the call for help would be heard.

On the basis that relay packages would be carried piggyback by enough satellites to ensure there would always be more than one in receiving distance, a continuous alert network quickly developed, and a wide range of weather, navigation, and Earth-resource monitoring satellites began to carry these life-saving beacon transponders. The first SARSAT transponder was flown in September 1982. Within five years ground stations had been built in 11 countries and more than 750 lives had been saved. The potential was obvious and NASA soon developed a **tiny emergency transmitter** that weighs only 1.5lb (680 g) for personal use. In emergencies the transmitter is activated, and a satellite somewhere in space picks up the signal and informs a ground station that someone is in need of help.

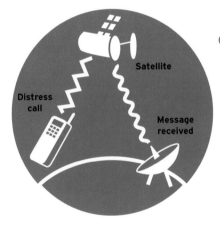

Satellite

Distress call

Message received

◉ A rescue call begins when a transmitter is switched on, sending a call for help to the nearest available satellite, which in turn relays the message to a rescue center.

A WATCH WITH A DIFFERENCE

Miniaturized technology allows an emergency transmitter to be carried inside this Breitling watch. If danger threatens or an accident befalls the wearer while he or she is out of communication range for conventional radios, the transmitter can be activated by hand.

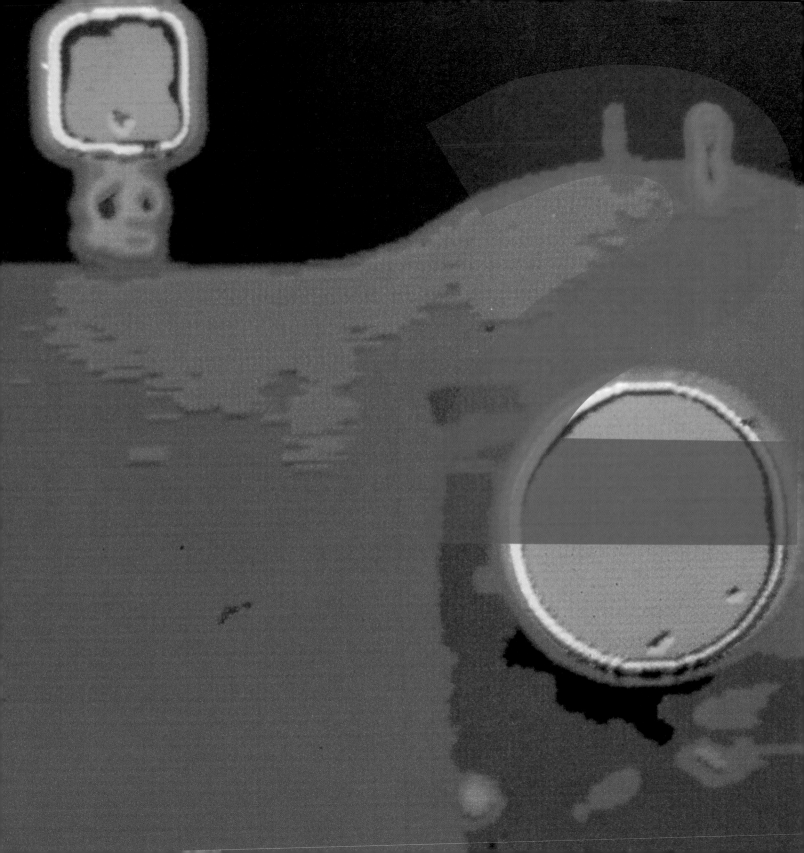

ENERGY, ENVIRONMENT, AND RESOURCE MANAGEMENT

NASA's SPACE EXPLORATION PROGRAM HAS REVEALED THE UNIQUENESS OF **EARTH**, AND HOW THE PLANET'S FINITE RESOURCES ARE A VALUABLE ASSET FOR ALL LIVING THINGS. **THAT** KNOWLEDGE HAS STIMULATED NEW WAYS OF LOOKING AFTER OUR ENVIRONMENT.

Mapping our world and listing its finite resources is the first step toward living in harmony with a fragile planet. Satellites directly benefit millions of people every day by monitoring crops for signs of disease (see p. 40), while techniques first developed for computer enhancement of pictures from Mars improve the quality of images of Earth (see p. 44). When it comes to developing our cities, NASA has helped urban planning through the use of infrared images (see p. 52), and human waste is now purified by plants, using a process developed by NASA (see p. 56).

Infrared sensors that are used to "see" the energy that is invisible to our eyes are helping archaeologists uncover information about the lives and civilizations of our ancestors. In addition to showing the sites of long-forgotten cities, the technology is sensitive enough to pinpoint buried pottery that dates back to prehistory.

SATELLITES LOOK INTO OUR PAST

The human eye does not see very much. If the spectrum of all energy radiated from objects on earth and in space were laid out in a line around the Earth, the portion of light visible to our eyes would occupy a band about the width of a pencil. On one side of this band, the energy would go into the ultraviolet, and on the other it would go into the infrared, neither of which can be seen by the naked eye. NASA scientists developed **infrared sensors** that react to the heat from an object or material and can "see" into the invisible areas of the spectrum. This technology is used in satellite-based surveying equipment in order to locate mineral resources, measure crop blight, and to conduct inventories of land resources. Archaeologists have also used this technology to look into the past. When soil, thin layers of volcanic ash, or vegetation cover ancient buildings they are almost impossible to detect with the unaided eye. However, when satellites or high-flying aircraft use instruments that "see" the surface in several bands of the infrared, the subsurface structures stand out like three-dimensional images.

In the mid-1970s inhabitants of the area around Lake Arenal in Costa Rica found artefacts from an **ancient civilization**, prompting the University of Colorado and the National Geographic Society to go in search of an established

⬆ **This magnified image of the Lake Arenal area in Costa Rica helped archaeologists to pinpoint the location of an ancient civilization.**

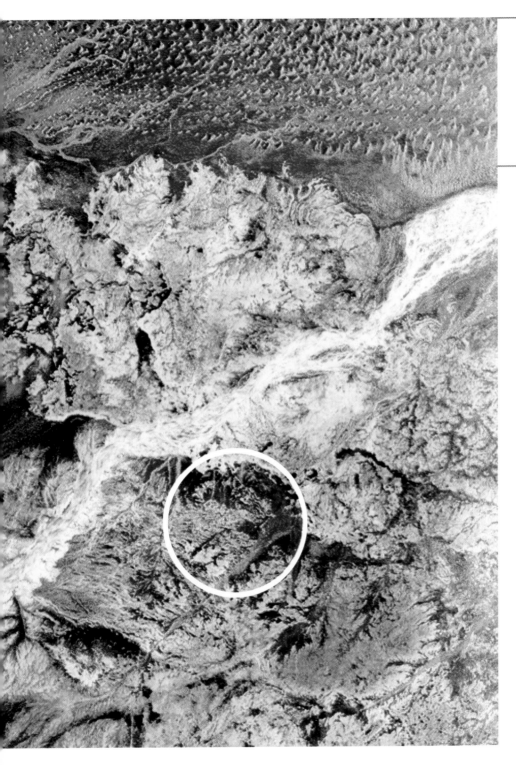

community. Very little evidence of dwellings was found on the ground, but in the 1980s **infrared satellite images** were used, together with data from a NASA remote-sensing aircraft, to pinpoint 62 sites, including villages, cemeteries, buildings, and tombs. The satellite pictures even revealed buried pathways, springs, and quarries.

Satellite images have also revealed sites in New Mexico that are among the oldest evidence of human occupation in North America. In Kenya, East Africa, anthropologist Richard Leakey used infrared images to map structures 200,000 years old. The 5,000-year-old Mesapotamian city of Ubar in Iraq was discovered using similar techniques by a team led by American George R. Hedges and British explorer Sir Ranulph Fiennes. Hedges and Fiennes used remote sensing and **satellite navigation data** to uncover the remains of a magnificent city with walls more than 10 feet (3 m) high and towers enclosing a busy trading area.

Everyday, people need information about soil use, land management, and where to find the nearest source of water. Now, by integrating remote sensing data and map data and merging them onto a single screen, computer operators can get the results quickly enough to make the information work productively.

IN 1973 SKYLAB PIONEERS THE USE OF EARTH-RESOURCE SENSORS

LAND MAPS
THAT SAVE LIVES

⬆ **Monitoring the Earth by satellite to detect subsurface lakes or wells. Scientists can now locate scarce water resources quickly and accurately using tailor-made computer software.**

In 1973 NASA used leftover spacecraft and spare rocket stages to launch the first US space station. It was called Skylab and was manned by three teams of three astronauts who lived on the station for up to three months at a time. Skylab pioneered research into the information potential of orbital platforms. Remote **sensing of Earth's resources** was particularly high on Skylab's list of priorities.

Using large scanning instruments and cameras, the Skylab crew were able to respond to an urgent call from a United Nations aid agency requesting information about a drought-stricken part of the southern Sahara, where nomads were searching for fresh water. The Skylab crew used multispectral photography to reveal where subsurface water could be found. However, there was so much information to search through, and the sifting process took so long that many nomads died before the information could reach them. Aware of the need for rapid access to the right sort of information, a group of NASA scientists worked with industry to adapt products originally developed to help speed up information in the space research centers. They formed a company called Delta Data Systems Inc., based at Picayune, Mississippi, to develop and worked on a **rapid-response computer program** that could be used to process remote-sensing data.

The Atlas Remote-sensing and Information System (ARIS) was the result of Delta Data System's work. The ARIS software program takes digital soil data and links it with topographic maps to generate what scientists call **land-use maps**. Delta Data Systems have built on this to make major changes to the way small- and medium-sized operations can gain access to information without having to wait months for laboratories to process the data.

The key to the success of ARIS is that instead of trying to cram every bit of information and data into the program, the software is built to provide just the **right level of information**. Rather than flooding the operator with data, the system gets straight to the heart of the matter.

LIFE OR DEATH

Nomadic people on desert margins live on the fringe of survival. Locating resources is crucial, and satellite data can provide vital information about food and water when drought conditions strike.

IMAGE-PROCESSING CROP INVENTORIES

NASA FOCUSES ITS ATTENTION ON VEGETATION AND LAND USE

The planet's resources are viewed on a daily basis from near-Earth orbit by satellites sent on routine missions to catalog forests, wetlands and vegetation. The need to maintain a comprehensive watch on selected areas of the planet is backed up by special aircraft fitted with remote-sensing equipment.

Remote-sensing of the Earth and its precious resources is vital to the health and welfare of the planet. Since the first rockets took pictures of the receding Earth, the beauty of our home planet amid the grayness of barren worlds that surround the sun has been a catalyst for **environmental concern**. During the 1970s NASA shifted the focus of its attention from the Moon to the planet Earth.

Information about the health and vitality of vegetation and crops is obtained by analysis of sunlight reflected from the ground. Infrared images also reveal subtle changes that indicate disease or poor growth in plant life. Through **infrared mapping,** scientists can pick out individual trees in an orchard, enabling growers to give attention where needed. Urban planners use resource pictures to help maintain an acceptable balance in the growth of cities, and farmers consult image libraries to keep watch on the health of crops. However, whether from space or airborne sensors, the data is frequently compromised by changes to lighting angles and brightness levels—known as differential lighting.

Positive Systems of Whitefish, Montana, has joined forces with NASA's Stennis Space Center in Mississippi to tackle the problem of differential lighting, a characteristic known as "bidirectional reflection." By taking digital images and merging them together using a **new computer program** based on NASA algorithms, Positive Systems can apply corrections to remove variations and restore the image. It can be compared with other pictures taken at different times of day under different lighting conditions. Computer processing enhances the image and makes it possible to reveal objects smaller than 3ft 3in (1m) across in pictures sent to Earth by satellite or aircraft.

Under a special license granted by NASA, the company is marketing the process to frequent users of **remote-sensing imagery**. Positive Systems' clients include town planners, the US Forest Service and National Park Service. Food producers and wine growers also make use of this innovative technology. All of these users need highly accurate and refined data, and to supply this Positive Systems uses a special airborne camera with four sensors that capture images through blue, green, red, and near-infrared color filters. This enables aircraft to make precise surveys in almost all weather conditions.

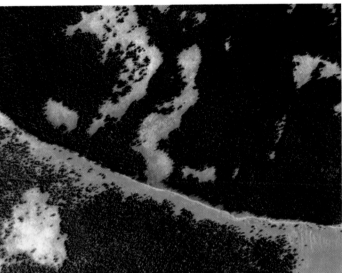

GETTING CLEAR VIEWS

Obtaining good-quality images of the same area over a period of several months is an important step in efficient land use. Using NASA algorithms, computers can then analyze the images taken under different lighting conditions, and generate comparative data about crop yields and diseases.

For dolphins and porpoises, the most dangerous fishing tool is the gillnet. This deadly device is designed to sink to the sea floor and catch bottom-dwelling fish that populate inshore waters. By attaching "bleepers" to these nets, fishermen can now warn small cetaceans off, saving them from the danger of becoming entangled.

DOLPHIN-UNFRIENDLY FISHING

Trapped in the strong mesh of a gillnet several miles long, and unable to return to the surface, this striped dolphin has drowned in the Atlantic Ocean.

APOLLO SPACE CAPSULE CARRIES A BEACON TO AID PRECISE LOCATION

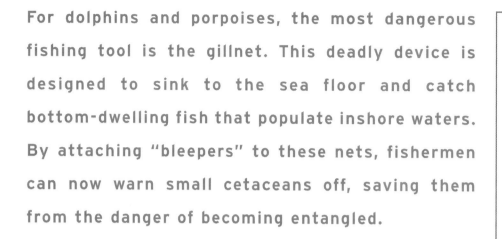

SAFEGUARDING DOLPHINS

This tiny device, the size of a small fish, was developed from technology invented for precise location of spacecraft and aircraft. Attached to a gillnet, the device gives out an audible warning at a frequency regarded as a danger signal by dolphins and porpoises.

Back in the 1960s NASA faced the challenge of devising aids for locating reentering spacecraft that had plummeted back through the atmosphere. With two-thirds of the Earth's surface covered by water, NASA needed a device that would not only withstand the shock of a high-velocity impact but could also send out a signal for several days from deep water. To locate these watery payloads, engineers at the NASA Langley Research Center designed a shock-resistant device that would send **multi-directional signals** for hours on end, giving recovery teams time to find and retrieve the errant object.

Dukane Corporation's Seacom Division of St. Charles, Illinois, made a business out of selling **audible beacons**. These devices were initially used for locating the black boxes on airliners that record crash information, as markers for underwater survey or exploration sites, and as location buoys for hazardous waste deposits. Dukane —working in collaboration with the University of New Hampshire's Department of Ocean Engineering, the Woods Hole Oceanographic Institute, and the New England Aquarium—subsequently developed a special bleeper which sends out a "pinging" sound to warn dolphins and porpoises of nearby danger.

In cooperation with Burnett Electronics of San Diego, Dukane produced a bleeper called NetMark 2000, a device less than 6.6 in (17 cm) long and 2.3 in (6 cm) in diameter. Netmark 2000 is attached to each sink-gillnet used by offshore fishermen and is rugged enough for deck handling. It is also capable of sending out a signal for **over 35 days** on the power of just four AA batteries. NetMark 2000 emits an audible signal every four seconds over a range of 328ft (100 m) from depths of up to 100 fathoms (182 m).

Dukane's NetMark 2000 has successfully reconciled environmental priorities with the needs of fishing communities whose waters are frequently visited by porpoises. With **bleepers attached to gillnets** porpoises are given adequate warning of places where nets rest on the bottom. With over 100,000 bleepers sold, Dukane is working on devices that will protect a wide variety of wildlife in and out of the water.

VIEWING THE EARTH— AND YOU

ELECTRONIC CLEANING IMPROVES PICTURES FROM MARS

Image enhancement has provided stunning pictures of other worlds and refined views of our own planet's finite resources. It has also given improved pictures from X-ray scans of the body, allowed physicians to "see" inside the brain of a living person and aided the classification of chromosomes in the human body.

When NASA sent the first probes to the Moon in the early 1960s, engineers were aware that signals received back would be contaminated by **electronic interference** between Earth and the spacecraft. Utilizing technology that predated the Space Age, NASA perfected a system of "scrubbing" the signal to remove interference. Image enhancement also helped to compensate for poor lighting angles, bad contrast levels, and fuzzy edges to blurred images. By the 1980s, when more sophisticated images were sent back from Jupiter and Saturn by Voyager spacecraft, the technology of image enhancement had been perfected. Now the system can sort out any imperfection without putting in things that are not there or guessing at the darkened sections. The results are stunning, as demonstrated by the **crystal-sharp images** sent back from the surface of Mars.

Long before NASA's image-enhancement technique was perfected, the system was applied to photographs of the Earth sent back from space. Useful images of the planet's mineral and agricultural resources were soon created from dark, fuzzy pictures. The enhancement process was also used to perform **complex management of digital signals** sent to Earth from satellites in space. UNYSIS Defense Systems of Camarillo, California, developed an image-processing software package that found customers from the US Department of Defense to the US Postal Service.

ERTS 1
ASCS
HOLT COUNTY, NEBRASKA

LEGEND: 6 CLASSES
RED	CORN
L. GREEN	HAY GRASS
D. GREEN	PASTURE
ORANGE	SUNFLOWER
PURPLE	ALFALFA
L. PURPLE	ALFALFA AND GRASS
BROWN	BARE SOIL

SUPERVISED COMPUTER CLASSIFICATION MAP   7/30/72

RED	FIELD CORN
YELLOW	POP CORN
ORANGE	SUNFLOWER
GREEN	ALFALFA AND GRASS
D. GREEN	PASTURE
BLUE	BROOME
PURPLE	ALFALFA
BROWN	BARE SOIL

TEMPORAL OF 7/30/72 AND 8/16/72

CLEANING THE IMAGE

This false-color image (left) of Holt County, Nebraska, was taken by multispectral satellite cameras that "see" the ground at infrared wavelengths. The computer-enhanced version of the image (below left) reveals individual fields and crops.

The UNYSIS software program can filter noise, enhance contrast levels, adjust the apparent surface lighting level, and **integrate separate images** by blending colors. The program can even stretch contour levels into three-dimensional images to give photoanalysts and interpreters a clearer picture of surface elevation. From this work, and that of many other teams, NASA's Jet Propulsion Laboratory has funded a research program aimed at improving image-enhancement techniques for medical applications. The next time you have an X-ray, the radiographer will have a clearer view of what is going on inside your body thanks to technology developed for Moon probes and Mars landers.

FURTHER USES

- Image-enhanced brain scans
- Improved scans of engine parts
- Instant land-contour maps
- 3D body imaging

A computerized solar-powered water heating system, which is connected to a voice synthesizer and answers back when the user asks for information, sounds like pure science fiction. But such a device does exist and was developed directly from research into how astronauts can survive months orbiting the Earth.

THE WATER HEATER THAT TALKS BACK

SOLAR-POWERED HEATING BRINGS HOT WATER TO SKYLAB ASTRONAUTS

In the early 1960s when NASA first planned **Apollo expeditions** to put astronauts on the Moon, argument raged on how best to achieve that amazing objective within the remaining years of the decade. New technology had to be developed quickly and answers had to be found to questions about how humans react to space flight. After all, not one American had so much as orbited the Earth when the Apollo goal was announced. Among all the other concerns that were thrust into the forefront of the expanding space program, a great deal of research had to be conducted in a short time to discover how to generate heat efficiently and safely to protect people in a spacecraft from the extreme cold of space.

Many rehearsals with Apollo equipment were needed and it was widely believed that the landing itself would probably only take place after several canceled or aborted attempts. In reality the goal was achieved with lightning speed. Only four flights with astronauts in Apollo spacecraft were required before the landing was successfully accomplished at the first attempt. NASA used the hardware left over from the rehearsals to put together an earth-orbiting space station called **Skylab,** which was launched into orbit in May **1973,** just six months after the last Moon flight.

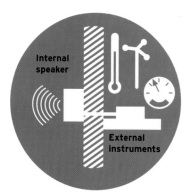

Internal speaker

External instruments

⬆ **Freesource uses external monitoring instruments and informs the operator about weather conditions.**

HOT WATER FROM THE YARD

The solar panels that provide the heating for the Freesource and Energy Garden units can be placed almost anywhere, from the roof of the house to the wide-open space of the backyard.

Skylab would provide the springboard for many technological developments. Exemplar, Inc., of Hickory, North Carolina, designed a solar heating system, known as the **Energy Garden,** based on the temperature control system on Skylab. The Energy Garden can deliver up to 70 percent of the hot water needed for a family of four, and with an amplifier added it can supply 90 percent. The unit is not only a water heater, it is also a miniature weather station, linking conditions inside with those outside.

In the late 1980s Exemplar added a more efficient second-generation unit called **Freesource.** The new system is based around a three-way heat exchanger that freezes and melts to release or absorb heat according to external and internal sensors. Freesource is a direct copy of the Skylab system and provides efficient, maintenance-free heat storage in a compact package. The glazing material that insulates the heater comes from thermal insulation developed for the Apollo Moon-landing vehicle.

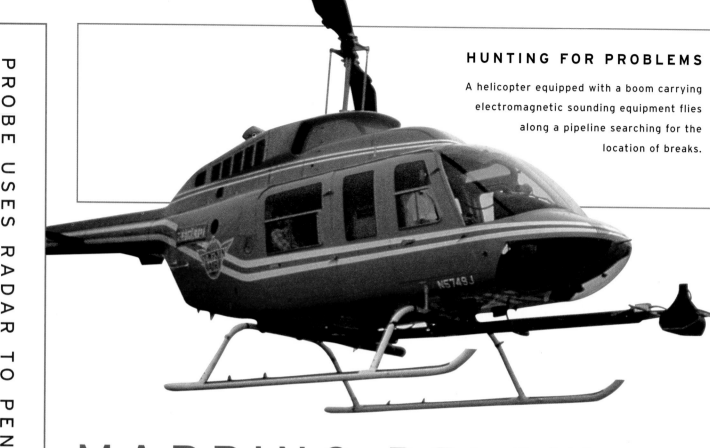

HUNTING FOR PROBLEMS

A helicopter equipped with a boom carrying electromagnetic sounding equipment flies along a pipeline searching for the location of breaks.

MAPPING BELOW THE SURFACE

In remote areas, where the Earth's surface is unaffected by towns or cities for hundreds of miles, it is extremely difficult to locate leaks in pipelines. Pinpointing a leak with a land-based vehicle can take days or even weeks. However, a NASA-developed geologic sounder mounted on a single aircraft can do the same job in a matter of hours.

In the early days of spaceflight, **scientific equipment** had to be light and small. Even on journeys that could last months or years, rockets could lift only a few hundred pounds. Over time, engineers and scientists throughout the United States worked with NASA to perfect sensors and power systems that would gather information for transmission to Earth on radio signals so weak they would not even be strong enough to power an electric lightbulb.

In the 1970s it took a long time to get to the planets, so scientists had to make do with fleeting glimpses of these other worlds from a fast flyby. Engineers and scientists worked together with NASA to devise ways of getting information from the surface of these worlds back to Earth without having to land on them. Research into **electromagnetic sounding** showed that it was possible to create a device that could scan and remotely probe the surface and peer beneath the rock and soil to discover underlying strata.

NASA's geologic sounder has found applications on the Moon and in spacecraft destined for deep space. Its most beneficial application to everyday life has been aboard a low-flying helicopter operated by AirBorne Pipeline Services, Inc. Boom-mounted sensors attached to the helicopter provide input for electromagnetic sounding systems that produce computer-enhanced views of subsurface structures. Much of the research into mounting the sounder on a helicopter came from NASA's Ames Research Center. Airborne Pipeline Services operate a specific type of geologic sounder— the Cathodic Protection Monitoring System—capable of obtaining **accurate geological surveys** to a depth of 650 ft (200 m). It has been used in the US for locating and mapping coal seams, mineral-bearing deposits, and boulder or bedrock formations in goldfields. Perhaps its most useful application has been in peering beneath layers of asphalt to map underground pipelines, cables, and buried power lines. The system can also be used to seek out subsurface pollutants and to monitor oil and gas conduits for early signs of cracking and leakage.

⊍ An image created by the electromagnetic sounder allows scientists to check the integrity of a pipeline running almost 3 feet (1 m) below the ground.

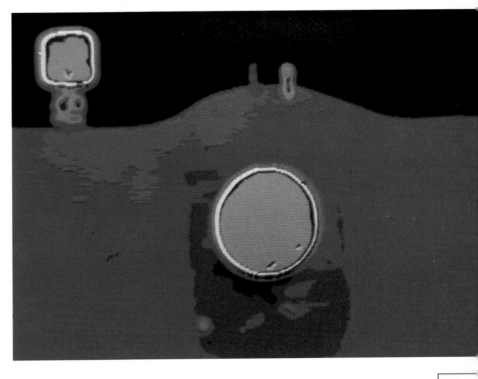

Pollution control is crucial to preserving the Earth in good condition for future generations. One way to begin the big clean-up is to set pollution limits at achievable levels and then lower them year by year. A system for managing resources developed by NASA scientists offers a means for companies to benefit from the successful control of waste.

POLLUTION CONTROL THROUGH NASA-STYLE EXCHANGE

In Southern California a system of credits is awarded to each industrial company. One credit allows the company to dump a certain amount of toxic waste into the atmosphere, but when that credit is used up the company must stop. The effect of this system is to cap the total amount of pollution that can be legally produced in a given area. To survive, each company needs to balance the amount of toxic pollution it produces over the year. If a company is effective at reducing emissions, its credits can be sold to less successful ones at a profit. The system gives companies an incentive to clean up since they can actually **make money by reducing toxic waste**.

The management of this giant budgeting job was daunting. Local government officials did not believe it could work. However, they changed their minds when they heard of a similar problem solved in a unique way by a team of NASA engineers working on the Cassini spacecraft launched toward Saturn in October 1997. Managed by the Jet Propulsion Laboratory (JPL), the program faced an early challenge in allocating spacecraft resources, and the Cassini Resource Exchange (CRE) was developed to

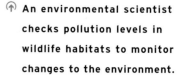

An environmental scientist checks pollution levels in wildlife habitats to monitor changes to the environment.

support the process. Operating like a commodities exchange, CRE **rationed spacecraft resources** by allowing, for instance, scientists who wanted more power to operate their instruments to trade mass or computer time with another investigator for additional power. CRE utilized principles of exchange, inducing each of the instrument teams to divulge their resources and trade among themselves, creating a sort of "bartered exchange" system to arrive at fair market "prices." In April 1995, Scholtz and Associates of California took this system and established the Automated Credit Exchange (ACE) to manage pollution credits for the companies in the region. The system now enables companies to trade what are, in effect, emissions licenses.

ENCOURAGING EFFICIENCY

Power generation is a major source of atmospheric pollution. The exchange system that allows companies to sell their pollution credits enables them to profit from increasing their efficiency and thereby reducing pollution.

On our crowded planet, selecting the right areas to develop into new towns is a vital part of caring for the environment. Satellites, with their all-seeing view of the planet, can help assess the full impact of a potential development, capturing the whole site in a single photograph.

SATELLITES ROUTINELY SCAN THE ENTIRE SURFACE OF THE EARTH

INFRARED MAPPING

Detailed, information-packed infrared satellite images like these help planners minimize the environmental impact of development in urban and rural areas.

INFRARED URBAN PLANNING

The way urban planners balance the needs of people and the environment has long been a matter of concern for space agencies. Egypt was one of the first countries to recognize that images from space could help, and has been working with space agencies for decades to obtain **infrared images that help plan urban and rural development**. NASA began launching remote-sensing satellites in 1972, and for six years the Egyptian Ministry of Development conducted a study to see how the expansion of Cairo's city limits was affecting the Nile River. With infrared images the different areas of vegetation, wetland, urban development, and desert were clearly visible, and this ability to see the balance between the categories provided information for better regional planning.

The Egyptian study was funded by the US Agency for International Development and the resultant data was given to Planning and Development Collaborative International (PADCO), based in Washington, D.C. Assisted by scientists from the University of New Mexico at Albuquerque, New Mexico, PADCO generated survey maps and area measurements derived entirely from the images sent back to earth by LANDSAT 1, NASA's first remote-sensing satellite. Computer-processed data gave a clearer view of the situation and built a picture of how regional planners could serve the needs of Cairo's burgeoning population while **caring for the environment** and preserving agricultural land.

Over its six years the Egyptian study built a coherent picture of how the ordinary life of everyday people was impacting on the environment. The study also revealed how the work and practices of local businesses was affecting the balance between rural and urban development. Due to an almost total lack of aerial survey photographs, LANDSAT was the only means of getting this **dramatic overview of an entire country**—it was also less expensive than having continuous strips of photographs taken by aircraft over the same period. This example and the lessons learned from it spawned an entire industry, and remote-sensing satellites were launched by Japan, China, and France, among others.

⬆ **Viewed from space, the use of slash and burn techniques in the Amazon jungle are seen to be destroying the valuable rain forest.**

ECOSYSTEMS FOR ENVIRONMENTAL PLANNING

MISSIONS TO OTHER PLANETS
STUDY ATMOSPHERIC SYSTEMS

In a world in which living things are dependent on a delicate balance between environmental extremes, all life is vulnerable. An ecology display developed from science instruments aboard NASA satellites has helped thousands of people to understand more about the Earth.

MINIATURE WORLD

This hermetically sealed glass bubble demonstrates, on a small scale, the self-sustaining ecosystem that has developed on Earth over millions of years.

Awareness of Earth's fragile environment began when weather satellites discovered a thin veil of ozone at extremely high altitudes. Ozone is a colorless gas with the odor of chlorine and is formed by the action of ultraviolet radiation or electrical discharges. It is highly dangerous when inhaled, but in the stratosphere, ozone shields the Earth with a spherical envelope that absorbs harmful ultraviolet rays from the Sun. The same satellites that discovered the ozone layer also measured its **gradual erosion by toxic products.** Much of this erosion has been found to be the product of human activity, although volcanoes and gases released from the Earth's crust have also contributed.

In the 1960s NASA discovered that Venus—a planet the same size as the Earth—had a dense carbon-dioxide atmosphere with a surface temperature of 752° F (400° C). The atmospheric conditions on Venus were caused by a phenomenon that scientists dubbed the "greenhouse effect." The depletion of the ozone layer around earth and the buildup of carbon dioxide in the atmosphere was soon attributed to the "greenhouse effect."

The Space Age has also helped enviromentalists gain a better understanding of Earth's fragile ecology through the development of **self-contained ecosystems**. NASA's research into the requirements for human exploration of the solar system resulted in the Jet Propulsion Laboratory in Pasadena, California, producing a balanced minature ecosystem capable of sustaining plant and animal life for several years.

Engineering and Research Associates, Inc., of Tucson, Arizona, adopted that technology for its EcoSphere: a sealed sphere 5 in (13 cm) in diameter, containing six tiny shrimps, several tufts of algae, and a clear "soup" of bacteria in filtered seawater. The system requires no cleaning and no feeding and is, in effect, a **model of the Earth's environment,** which only needs to be provided with light as its source of energy. The algae bask in the light and produce oxygen as they grow, and the shrimp breathe the oxygen while nibbling on the algae and the bacteria. The shrimp and the bacteria give off carbon dioxide—needed by the algae for photosynthesis and growth—while the bacteria break down shrimp waste into nutrients for the algae.

As a fascinating example of a sustainable ecosystem, the EcoSphere is a lesson to all that the Earth is a very fragile place and needs respect. As an educational tool for classroom work, it helps children realize that the ecological balance on the planet is a delicate one, but one that, with careful management, can be made to endure.

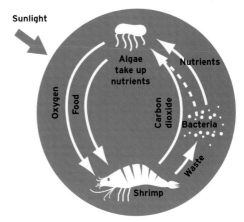

Sunlight

Algae take up nutrients

Nutrients

Oxygen

Food

Carbon dioxide

Bacteria

Waste

Shrimp

⬆ **The way in which all living things draw nourishment and energy from each other in a continuous cycle is called a "closed loop" ecosystem.**

FURTHER USES

○ Models for learning environmental lessons
○ Experiments into survivability
○ Aids for testing new cycles
○ Ways to discover different ecosystems

PLANTS THAT PURIFY SEWAGE

When you have a laboratory on your doorstep you can find some fascinating and practical uses for modern technology. Who would have forecast that a natural, safe, and economical way to treat waste sewage would emerge from NASA's rocket-engine test program?

A SPACECRAFT UNDERGOES VACUUM TESTS AT THE NSTL

CONTROLLING WASTE

Space centers use large quantities of energy and, as a result, create vast amounts of waste. NASA has developed a means of repurifying waste that utilizes the biological action of plants—in this case attractive water hyacinths.

In the 1960s NASA's new test facilities in the Bay St. Louis area of Mississippi became the National Space Technology Laboratories (NSTL). The expansion of this site and its test facilities meant that new utilities and waste plants were needed. In the early 1970s NSTL scientists discovered that glossy green **water hyacinths thrive on sewage**. By absorbing and digesting nutrients and minerals from waste water hyacinths are able to convert sewage effluents into clean water. The new system offered a means of purifying water at a fraction of the cost of a conventional sewage treatment facility. In addition, harvested hyacinths can be used as fertilizer, high-protein animal feed, or as a source of energy.

The first test of the plant purification system was carried out in a 40-acre (16-hectare) sewage lagoon at Bay St. Louis in 1975. Hyacinths were planted and the once-noxious lagoon quickly became a **clean water garden**. The concept rapidly attracted interest from towns and cities across the state of Mississipi. By the beginning of the 1980s many towns with populations of between 2,000 and 15,000 used water hyacinths as the primary method of treating waste water, while larger towns used this new aquaculture as a supplementary process to their main treatment operations. Local health authorities also consulted NSTL over the best way to scale up the system and invested in the waste-water treatment process.

Walt Disney World at Buena Vista in Florida installed a hyacinth aquaculture treatment facility in their Experimental Prototype Community of Tomorrow. This came to the attention of the city authorities in San Diego, California, who approached NSTL for advice about a large waste-water facility capable of handling 20,890 US gallons (95,000 liters) of sewage daily. San Diego soon built a treatment unit to process almost 879,600 US gallons (4,000,000 liters) of sewage a day with a special **reed-rock filter unit**. The San Diego unit is a hybrid aquatic plant/microbial filter combination which, unlike hyacinths, is able operate in cold climates.

⊕ **Protein-rich water hyacinths are cultivated for use as purifiers in sewage lagoons. They are grown in ponds (left) and gathered (center-left) for use in a reverse osmosis unit (center-right). Waste is passed through a pretreatment facility (above) before reaching the lagoon.**

57

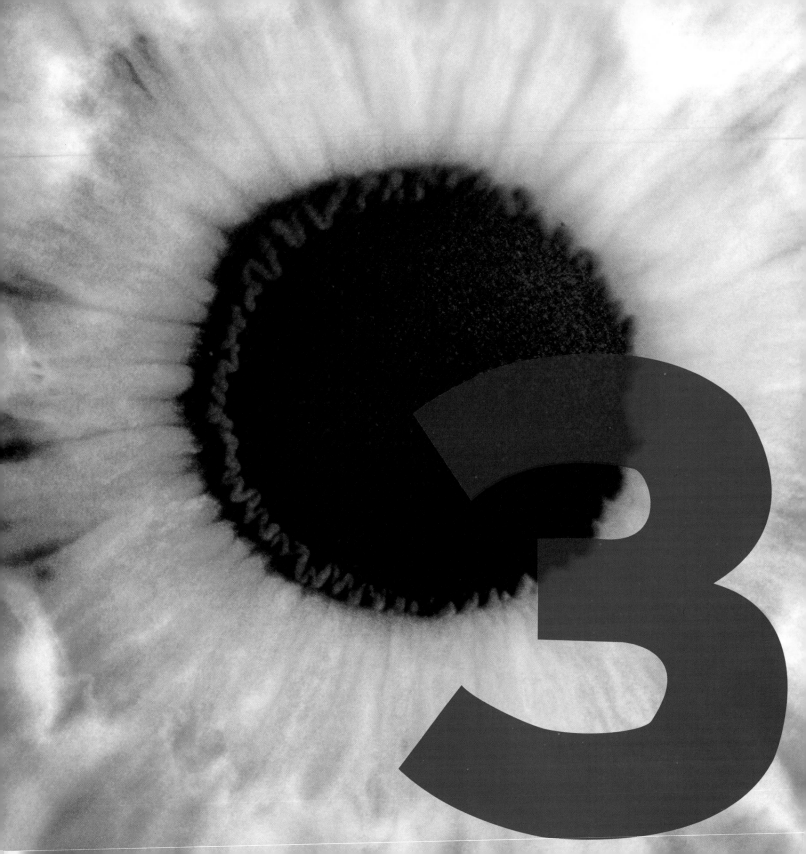

CONSUMER, HOME, RECREATION, AND ART

NASA TECHNOLOGY MAY HAVE BEEN DESIGNED TO DRIVE COMPLEX SPACECRAFT, BUT IT HAS INCREASINGLY FOUND ITS WAY INTO OUR HOMES. THE INVENTIONS OF THE SPACE RACE HAVE MADE A SURPRISING AND SIGNIFICANT CONTRIBUTION TO BOTH HUMAN COMFORT AND RECREATION.

Many of the problems solved by spacecraft and satellite designers have a common link with less exotic challenges on Earth. Engineers and scientists have put their minds together to focus on the needs of home dwellers, while other inventors have adapted Space-Age technology for creative art forms. Engineers have helped to make cleaner water (see p. 66), weather-resistant coatings (see p. 72), and helped to build faster yachts (see p. 80). Shuttle technology has aided the production of new jewelry (see p. 70), while athletes have seen their footwear benefit from the work of NASA (see p. 76).

When it comes to thermal insulation, astronauts have it made. A spacecraft is like a giant thermos, sealing the inhabitants from the hostile environment on the other side of the pressurized cabin wall. Space-Age insulating technology is now being employed on Earth to save money, enhance safety, and increase protection against extremes of temperature.

THERMAL INSULATION FOR HOMES

THERMAL BLANKETS WRAP THE SHUTTLE FLIGHT DECK

Spacecraft can experience temperatures as high as 212°F (100°C), but inside the living quarters the crew work at a comfortable 70°F (21°C). Even more extreme temperatures are found inside rocket stages, where **hydrogen is stored for fuel** at 422°F (217°C). The technology designed by NASA for the insulation of rocket tanks and spacecraft linings led to the development of a commercial product which has proved a huge success. Radiant R is a lightweight fabric designed as attic or ceiling insulation and is an innovation of the Buckeye Radiant Barrier company of Dayton, Ohio. The material has two layers of 99 percent pure aluminum separated by a single layer of polypropylene, which acts as a thermal break, and a nylon grid, which provides strength. Radiant R is perforated to allow moisture to escape and it controls radiant heat (that flows through the air between objects), conducted heat (that moves from one solid surface to another) and convected heat (that rises in the air between structures).

The Buckeye Radiant Barrier company claims their invention is **95 percent effective** in the management of all radiant energy. People with Radiant R in their houses say it saves 20 percent on fuel bills. There are other uses for this technology, for example, when it was applied to a gas-fired boiler room in a school, the room above became inhabitable as a classroom for the first time. One company installed Radiant R around their

A wide variety of insulation has been produced for home and domestic use from thermal materials originally developed for space.

shrink-wrap oven and found that it cooled the surrounding air, which improved working conditions, while also reducing heating costs. Quantum International Corporation of Washington is another organization that has benefitted from NASA's insulation technology. Quantum produce a sandwich board called Super Q that works as a thermal insulator. Super Q is made from flat sheets which are stiffened by an **expanded polystyrene core** and strengthened by steel wire. Panels can be used for walls, floors, ceilings, automobiles, and refrigerated delivery trucks, as well as inner wall barriers for new homes. For stock car drivers, who have to survive temperatures as high as 160°F (70°C) during a race, BSR Products, Inc., of North Carolina, manufacture a material that is incorporated into the cars for temperature control around the racetrack.

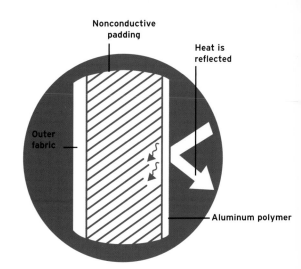

Nonconductive padding

Heat is reflected

Outer fabric

Aluminum polymer

KEEPING DRIVERS COOL

Strong, lightweight thermal protection is vital for both comfort and safety In motorsports where high temperatures are generated close to fuel pipes and drivers.

⊕ **To prevent a transfer of heat or cold, thermal blankets sandwich non-conductive material between sheets.**

GOLD-PLATED VISOR PROTECTS AN ASTRONAUT AGAINST THE SUN

The technology that protects an astronaut's eyes from the unfiltered rays of the Sun has led to an artform that graces the homes of world-famous actors and actresses. The link between the science found in an astronaut's sun visor and a prized Hollywood possession is literally out of this world.

NEW BEAUTY

A myriad of shapes and textures, illuminated by subtle shifts in light and shade, bring a new experience in optical art, which combines established sculpture techniques with Space-Age technology.

ART FORMS FROM OPTICAL GLASS

⬆ **Dichroic glass has become the new art medium for adding dimension to flat forms, reflecting and absorbing light in a variety of wavelengths.**

The unlikely link between astronaut technology and aesthetics was provided by Murray Schwartz, an aerospace engineer who left the high-tech world of satellites and lasers to satisfy his passion for colored glass. Schwartz's particular interest was in "dichroic," or two-colored, glass. Dichroic glass, which is commonly used in the optical industry, consists of very thin films of metal oxides that are vacuum-deposited on the glass surface. A key feature of this type of glass is that it changes **the color of a light beam** when reflected from its surface. When light falls on the glass, some colors in the spectrum pass through, while the colors we see are reflected.

Dichroic glass was developed in the 1950s and 1960s, and was used for coating light-sensitive screens on satellites and spacecraft. It was also famously used on the faceplates of Moon-walking astronauts. On Earth the glare of the sun's visible light is dramatically reduced by the dense atmosphere, but on the airless Moon, human eyes are vulnerable to the Sun's harsh rays. Without protection astronauts would be blinded, so scientists turned to dichroic glass to guard against harmful rays. **Ultra-thin deposits of gold** allow the astronauts to see out, but most visible light gets reflected back away from the faceplate.

With dichroic glass there is a direct relation between the type of thin metal deposits applied to the glass and the colors that are reflected back. Each material has a very **narrow color band,** and sharp distinctions between different colors are easy to achieve by controlling the variety of metal deposited on the glass. The intensity of reflection is determined by the thickness of the deposit, measured in thousandths of a millimeter.

Murray Schwartz was impressed by the beauty of dichroic glass, and in 1971 set up a business to sell it as jewelry and works of art. At first Schwartz and his daughter Regina Standel sold dichroic glass art from the **back of a traveling van**, but he later set up a studio called Kroma, Inc., in Venice, California. In 1985 Schwartz met and married fellow glass expert Rupama Schwartz, and they moved to Sante Fe, New Mexico. Now, the venture that started as a mobile store has blossomed into a successful business.

63

HEAT PIPES IMPROVE THERMAL CONTROL

If you were to put one hand in a hot oven and the other in a deep freeze, you would feel the extremes of temperature that satellites and space vehicles experience. In space, heat-pipe technology is used to counter the effects of extreme temperatures and the same technology can now be found in many Earthbound applications—including refrigerators, computers, and restaurant appliances.

NASA worked with the New Mexico-based Los Alamos Laboratory to find a solution to the problems of exposing sensitive electronic equipment to extremes of temperature. The result of this collaboration was the heat pipe, a simple **heat-transfer mechanism**. The heat pipe is a tubular device in which a working fluid alternately evaporates and condenses, transferring heat from one region of the tube to another without any form of external help. The low-temperature fluid is pushed to the hot end of the pipe, where it evaporates, absorbs heat, and flows back to the cool end of the pipe, where it condenses. The system keeps itself in balance through this alternating expansion and contraction from low density to high density according to temperature. In satellites and spacecraft the pipe is routed through areas where heat is produced, and in large vehicles, such as the Shuttle, it delivers hot fluid to external radiators.

During the 1980s NASA began a program to promote the commercial potential of heat-pipe applications. Burger King, Taco Bell, and other fast-food restaurants use heat pipes to control temperature and humidity, reduce condensation, and **increase comfort** for both customers and staff. Heat pipes can also help reduce power consumption and building deterioration. In tests against air conditioners, heat pipes showed a 38 percent increase in moisture removal, a 25–30 percent decrease in humidity, and a 17–29 percent cut in power consumption. As a result of such trials Taco Bell instigated a program of heat-pipe installation in its restaurants in the southeastern United States.

Thermacore Inc. of Lancaster, Pennsylvania, has taken the principle of the heat pipe and developed micro-miniaturized units for computers. In order to restrict the temperature of a notebook computer's 8-watt central processing unit to a maximum 194°F (90°C), Thermacore developed a miniature heat-pipe system based on a powder metal wick. The wick transfers heat from the processor to an area where free air flows, thus saving battery power that would otherwise be needed for a fan. Thermacore is now working on the development of heat-pipe transfer systems for use in **personal computers**, as well as high-power systems in laboratories and research facilities. It is also hoped that heat-pipe technology will find uses in cellular telephones, camcorders, and other appliances that generate heat as part of their operating process.

KEEPING HEAT SAFE

Developed from technology designed to remove heat from delicate electronic components in spacecraft and satellites, heat-pipes lie at the heart of powerful computers of the future.

Contaminated water can produce a catalog of infections, diseases, and pollutants. In developing countries, water pollution can have devastating effects on local communities and even whole regions. One system used to combat the effects of contaminated water was developed as a weight-saving measure for expeditions to the Moon.

CLEANER WATER FROM SPACE TECHNOLOGY

When Apollo astronauts went to the Moon they drank water generated as a by-product of their rocket's electrical energy system. In order to save weight, engineers had designed Apollo with an innovative electricity generator which was fueled by supercold hydrogen and oxygen. As the spacecraft drifted through space the hydrogen and oxygen was consumed by a fuel cell that produced enough electricity to run Apollo for two weeks. The total weight of **Apollo's energy system** was much less than conventional batteries, and it was also more rugged and effective than solar cell panels. So hydrogen and oxygen not only served to produce electricity in the three fuel cells on board, but also came out the other end mixed together as water, which was then used for the spacecraft cooling system, for drinking, and for reconstituting freeze-dried food.

To purify the water from the fuel cell NASA developed a **tiny electrolytic water filter**. The new system worked so well that it was adopted for the Shuttle program. However, the demands on the Shuttle system were much greater due to the bigger vehicle (100 tons vs. 15 tons) and the greater number of crew (up to seven rather than three). In the Shuttle system the filter uses silver ions in concentrations of 50–100 parts per billion in order to get rid of bacteria and odors.

Increased environmental concerns during the 1970s led to a growth in demand for domestic water filters. These devices were usually based around a gauze filter. However, Paul "Mike" Pedersen—the founder and president of Western Water International, based in Forestville, Maryland—wanted something better and got in touch with NASA to develop its technology. Using the latest ion-based water-purification concepts, Pedersen produced a range of filters designed to **eliminate lead in the water systems** of both developed and developing countries. A wide range of Aquaspace Compound Filter Media units is now available throughout the world. Pedersen's filters have helped prevent sickness and disease, and have also raised standards of health and increased the value of water sources.

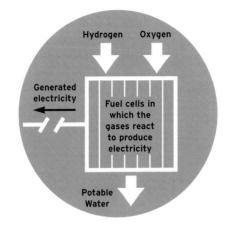

⤒ The electrolytic filter was developed from a NASA device to purify the water produced in fuel cells aboard spacecraft. The cells generate electricity from hydrogen and oxygen in a catalyst, and produce water as a by-product. The system thus provides two essential resources from one unit.

GIVING CLEAN WATER

Water purifiers designed for spacecraft are the foundations on which domestic purifiers have been developed.

At one company most of the employees work long hours thinking of ways to break into buildings, banks, computer rooms and top-secret government establishments—and devising means of confounding such attempts. They design and put together all forms of security systems used to secure classified or confidential papers, documents, and objects for NASA and other organizations.

COMPLEX MONITORING ENSURES HIGH SECURITY AT SPACE FACILITIES

KEEPING WATCH

Modern security systems for high-tech companies and government agencies include surveillance cameras, data banks, and computer-tracking of tagged personnel.

USER-FRIENDLY AUTOMATED SECURITY

The primary management tool used by Software House of Waltham, Massachusetts, has the cybernetic name of C*CURESystem 1 Plus and is a computer software-based system that organizes and manages complex corporate-security systems. While continuing to perform its design function, C*CURESystem 1 Plus interfaces with several other management reporting tools, among them systems such as the NASA CLIPS, developed at the Johnson Space Center. Using CLIPS, even inexperienced operators can move large volumes of data and solve **unusual or unique security problems**. As an empowering tool, it coordinates a variety of security devices: Card validation, alarm monitoring, closed-circuit TV systems, video badging, and biometric identification systems. CLIPS incorporates unique pathway access for specific individuals who are allowed entry to some locations but not others in the same building or facility. It can also program access-control hardware to lock or open doors according to the security map of the person concerned. CLIPS will even report back with an individual's route movements.

Software House's access to CLIPS was made via a NASA-sponsored system for making **complex computer software** programs available to businesses. Known as the Computer Software Management and Information Center (COSMIC), this NASA tool feeds such software programs directly to industry at a price that is a great deal less than it would cost companies to generate similar new programs for themselves.

COSMIC is located at the University of Georgia and is in essence a **library of computer programs** and software developed by the space agency for the use of businesses outside NASA. While COSMIC is an asset in itself, it is frequently used with other applications first developed as a result of NASA-based technology. In the case of Software House, programs that help provide user-friendly security systems are fed back into the COSMIC chain where other companies pick them up and apply them to their own unique products, starting the cycle all over again.

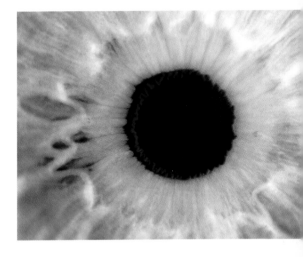

⬆ **The human iris is as personal to the individual as a fingerprint, and software-based security systems can now identify individuals by their eyes.**

JEWELRY DESIGN FROM THE SHUTTLE

HEAT-SINK TILES ON THE SHUTTLE DISSIPATE HEAT ON REENTRY

When jewelry designer and artist Alan Kasson gets to work on new creations he likes to keep a model of the Shuttle on his bench. It serves, he says, as a constant reminder of the special technology that gives him an edge in the production of exquisite works of art.

Kasson uses a jeweler's soldering torch to join precious metals or stones to their settings, however, with his torch operating at a seering **2,550-4,275°F** (1,400–1,800°C), delicate materials can easily become damaged. The solution to Kasson's problem was to adapt technology pioneered on the Shuttle. In order to protect the Shuttle orbiter as it returned, two insulating materials were used. Firstly, an ablative material that chars as it carries away the heat was employed and, secondly, heat-sink materials, which cause heat to dissipate rapidly from the surface and be absorbed slowly to the interior of the substance, were used. Ablative materials must be replaced after each flight, but heat-sink materials, which do not change their shape or distort, can be used over and over again.

Intricate items of jewelry can be held in pieces of insulation to dissipate the heat from the soldering torch while retaining a firm grip.

When NASA designed the Shuttle, the only available heat-sink material was made from extremely brittle silica fiber. In order to absorb the expansion and contraction of the orbiter's main structure the silica fiber material was applied as separate tiles bonded to the aluminum skin of the vehicle. There would be more than **34,000 separate tiles**, each specifically cut for its own location. The files were spaced with gaps between them that allowed the airframe to move without shattering the tiles. The heat dissipation process is so effective that the corners of tiles heated to 2,325°F (1,275°C) can be held by a bare hand while the interior glows red hot.

Alan Kasson soon realised that NASA's heat dissipation process was just what he needed to help hold his delicate metals while soldering them together. Kasson wrote to NASA's Johnson Space Center requesting some heat dissipation tiles for his work. Back came four tiles free of charge, but with a request that he test them out in the way he had described and report on their effectiveness. To Kasson the tiles seemed like soft firebricks, and the simplest way to use them was to push the item to be soldered into the fibrous tile. The results were stunning.

HEAT PROTECTION

Shuttle heat-protection tiles do such a good job of dissipating thermal energy that they can be held by their edges with unprotected fingers while their centers glow red-hot.

NASA'S LAUNCH PLATFORMS NEED PROTECTION FROM THE SALT AIR

For over a century, Liberty Enlightening the World has been buffeted by the winds and salt spray of Ellis Harbor, New York. Treasured as an icon of the free world, she overlooks the port where millions of immigrants first touched American soil during the 20th century.

PROTECTING LIBERTY

A high-tech ceramic coating known as IC 531 has been used to protect the surface of the Statue of Liberty from chemical pollution and the effects of weather. The coating prevents corrosion of the structure and reduces the need for restoration work.

WEATHER-RESISTANT COATINGS

↑ **Bridges and open-girder structures are now protected from the effects of weather by ceramic coatings using the chemical formula developed by NASA contracts for launch sites.**

FURTHER USES

○ Religious temples
○ Government buildings
○ Oil rigs
○ Piers and
  harbor facilities

A paint used to protect the Apollo launch pad has recently been used to give the Statue of Liberty a makeover. NASA scientists had to invent a new paint technology to produce a protective coating. The result was IC 531, a **quick-drying formula** that can withstand sudden temperature changes without cracking. IC 531 is designed to protect any iron or steel structure from corrosion, even in severe sea spray, wind, or fog.

The Apollo missions of 1967 to 1972, which took man to the Moon, used the most powerful rockets in the history of space exploration. When the Apollo rockets were ignited, it was not only the spacecraft but also the steel structure on the Cape Canaveral launchpad that needed protection from the resultant **thermal shock**. Unlike the Apollo spacecraft, though, the launchpad would have to be reused.

NASA scientists pushed forward ceramic chemistry to produce a paint that could protect the launchpad from **hot rocket exhausts**. The new coating would have to protect the steel structure from the effects of the nearby Atlantic Ocean while undergoing huge temperature change without cracking. In the early 1970s, NASA applied the first batch of their newly-designed **protective coating** to the Cape Canaveral launchpad. The scientists found that their quick-drying formula, now called IC 531, could protect the launchpad adequately using just one coat. Since 1982, IC 531 has been made under license by Inorganic Coatings, Inc., of Malvern, Pennsylvania, for use on more down-to-earth structures. Despite its uninspiring name, this paint has extended the lives of many international, landmarks—from San Francisco's Golden Gate Bridge to the Po Lin Buddha.

SMOOTHER JOURNEYS

NEW TECHNIQUES DAMPEN POWERFUL
ROCKET-ENGINE VIBRATIONS

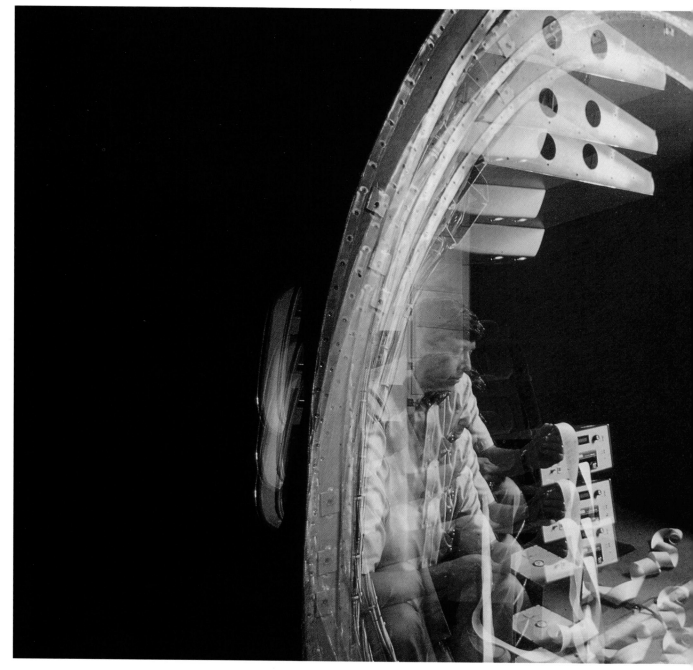

Vibrations can be destructive for machinery and dangerous for people. Annoying at best, at worst they can damage the spine and place great stress on the body's vital organs. In extreme examples, vibrations can set up a resonating pulse that interferes with the rhythm of the heart, sometimes with alarming consequences.

To land humans on the Moon, NASA had to build powerful engines that could lift 3,000 tons (3,048,000 kg) of rocket. The engines on the early lunar missions produced more energy than that generated by 500 fighter aircraft and consumed sufficient fuel to power an average car around the world more than 800 times. All that power was contained within a structure more than 49 feet (15 m) taller than the Statue of Liberty. When the engines ran at full power, the vast quantity of fluids consumed set up **vortex patterns** that created vibrations. As each part of the vehicle vibrated, or oscillated, at a specific frequency it stressed components and fluid lines. Sometimes vibrations shook apart delicate components and connectors. NASA likened these oscillations to the motion of a pogo-stick, a telescoping sprung pole designed to bounce up and down with vertical motion. In a program dubbed **"pogo-suppression,"** NASA engineers devised a way of injecting fluid into the propellant which caused the vehicle to vibrate at a slightly different and less harmful frequency.

NASA's Langley Research Center devised a way to take this vibration monitoring and apply it to the needs of industry. Soon the technology that stopped rockets from resonating was helping engineers precisely measure vibration and ride quality in seats designed for cars, boats, planes, and trains. Sensors and meters measured **motion harmonics** and their effect on the body, while different types of vibration were artificially programed into the machine to test the response to different frequencies. The information input into the machine provided the basis for the development of a computer model that translates subjective measurements into a single **discomfort index**. Sensors in the ride simulator measure vibration in five different directions, and microphones measure noise levels. The device was quickly taken up by Ford, Amtrak, and agencies responsible for land and air transportation.

BUMPY RIDE

A time-lapse camera catches the bone-jarring vibrations of a test rig designed to measure the ability of special devices developed to smooth out the ride and reduce discomfort.

Every hop, step, and jump you take when you wear an athletic shoe with an air pressure midsole involves an adaptation of NASA space technology.

COMFORT AND EFFICIENCY

Running shoes designed with the same kind of flexible bellows applied to spacesuits provide added comfort, and move more efficiently as the runner shifts his or her weight along the length of the shoe.

PRESSURIZED SPACESUIT PROVIDES PROTECTION AND FLEXIBILITY

SPACE-AGE SNEAKERS

Research and design teams at major sportswear manufacturers have long sought to develop an athletic shoe that would retain its **shock absorption, stability, and flexibility** over an extended lifetime. When aerospace engineer Al Gross of Lunar Tech Inc., Aspen, Colorado, was contracted to design such an advanced shoe he immediately investigated the potential of NASA technology. His approach was to find a way to eliminate the foam materials used in the midsole, because he felt it was these that were mainly responsible for the cushioning loss.

The clue to developing a super sneaker lay in the pressurized spacesuits used on NASA space missions. Pressurized suits are rigid but allow the astronaut a level of mobility via a "convolute system," which uses a series of **bellows in the joint areas** that expand and contract with every motion. By layering, or combining materials and varying the shape, size, and number of bellows, spacesuit designers can vary the flexibility of suit joints. Gross wanted to apply this concept to the athletic shoe. His task force created an **external pressurized shell** with horizontal bellows for cushioning and vertical columns for stability. By varying the shape, number, and thicknesses of shell materials, the designers fine-tuned the stiffness and the cushioning properties of the midsole.

Gross's design had developed quickly but was still in need of one more innovation. To ensure durability, a single part without weld lines or seams was necessary. Another NASA development was explored. On this occasion a blow molding process, which was first used to get impact resistance for the Apollo lunar helmet and visor, was examined. Blow molding allowed the manufacturers to reconfigure the compression chamber for different sports. In durability tests the compression chamber midsole was subjected to **stresses equivalent to 400 miles of running** and showed no signs of wear or structural fatigue. From this first step, the athletic shoe is now truly Space-Age, a piece of high-tech brilliance streamlined to provide speed and agility.

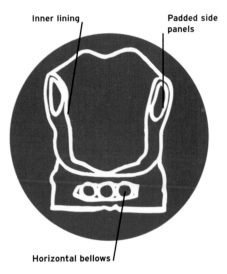

Inner lining

Padded side panels

Horizontal bellows

⬆ **Seen here in cross-section, the sole of the shoe developed by Lunar Tech has an external shell enclosing horizontal bellows that cushion the load.**

FURTHER USES

○ Scubadiving suits
○ Heavy protective clothing
○ Heavy-duty workshoes
○ Flexible tube seals

ADVANCED SOLAR WATER HEATING

SPACECRAFT HEAT SYSTEM WORKS ON TEMPERATURE AND PRESSURE

During the course of the 20th century the population of the world increased fourfold. As a result, the need for energy now threatens to outstrip available resources. The challenge of providing sufficient energy has galvanized scientists and engineers to make advances with ideas that challenge logic but promise solutions.

One scientist who has wrestled with the problem of finding alternative energy sources is Dr. Eldon Haines, formerly of the Jet Propulsion Laboratory (JPL) in Pasadena, California. JPL is the home of NASA's **unmanned planetary exploration** program, and is under contract to NASA to carry out the remote probing of other worlds. The relationship between CalTech, JPL, and NASA frequently brings scientists, engineers, and technologists together on unique encounters with common problems.

Dr. Haines was a specialist in "geyser pumping," the means by which heat and water can be made to move by varying temperature and pressure. Believing that the principle behind geyser pumping could be applied to make a solar heater, he reversed the two physical reactions and invented a device that would do the job with **no moving parts**. The result was a machine that combined passive and active systems to deliver up to 90 percent of the hot water used daily by an average family.

In 1977 Dr. Haines left JPL and completed development work on his Copper Cricket home water-heating device, utilizing shallow flow trays installed on the roof. By the late 1980s the product was ready to be marketed. Haines' system was extremely low maintenance and provided a **fully independent solar-heating unit** that was immune to frost and required no roof-mounted tanks. With two colleagues, Dr. Haines formed Sage Advance to put the idea to the public via a commercial enterprise. Next came Copper Dragon, a scaled-up unit designed for larger buildings and health centers. Copper Dragon was equipped with multiple shower units and washrooms, ideal for campsites, small work units, and business premises. Much of the work Dr. Haines had conducted at JPL was relevant to his ingenious and creative designs. Haines continued at Sage Advance but also returned to work with NASA as a consultant, putting back into the source of his inspiration ideas for future inventions.

GEYSER ON THE ROOF

Geyser-pumping solar water heaters are easy to assemble and install, separate panels being placed in the quantity required to meet the needs of the house. Each of the units is independent and self-contained, but they can be coupled together to operate as a single system.

⬆ Geysers move heat from subsurface mud to the atmosphere through steam, which acts like a pressure pump and expands. The advanced solar water heater works in the same way, using a balance between temperature and pressure.

⬇ The diaphragm in this pressure pump is operated by the flow of liquid, which slowly passes through a solar heater that warms it up.

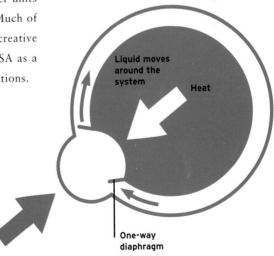

Liquid moves around the system

Heat

Heat

One-way diaphragm

"RIBLETS" FOR SPEED AND MEDALS

F-16 IS MODIFIED TO STUDY AIRFLOW ACROSS DIFFERENT WING SURFACES

What is the link between a shark, a NASA wind tunnel, and winning an Olympic medal? The answer is the search for a way to reduce aircraft fuel consumption by as much as 10 percent.

SECRET OF SPEED

Under the microscope an individual riblet looks like the upper surface of a seashell, but the separate grooves have the property of smoothing the flow of air, reducing drag, and increasing speed.

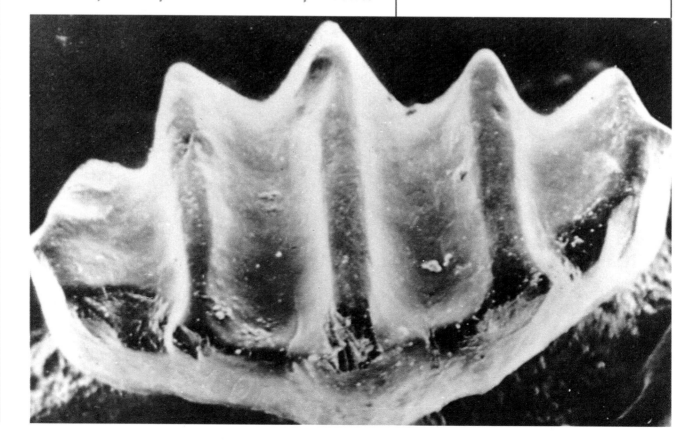

In the mid-1970s engineers at NASA's Langley Research Center began work on reducing skin friction (the "sticky" effect that air picks up as it passes across aircraft wings). A phenomenon called **"turbulence bursts"** causes most skin friction on wings. As air flows across the wing, it breaks up the turbulent flow and creates tiny bursts—violent eruptions on a very small scale—that produce most of the unwanted drag. In their research, engineers found that small, almost imperceptible, grooves on the surface of the aircraft can reduce skin friction by as much as 10 percent. V-shaped and with the angle pointing in the direction of flow, these grooves were dubbed "riblets" because their shape resembles minute ribs with a depth no greater than a scratch.

US aircraft manufacturers picked up on riblet technology, and it was soon discovered that the skin of **fast-swimming sharks** is covered with tiny riblets that increase the sharks' performance in the water. In 1984 the men's U.S. four-oar-with-coxswain team won a silver medal at the Olympic Games using a shell covered in adhesive riblet material. As a sidebar to the main research, it was a good way to promote the technology and get the attention of the aircraft makers. Three years later, in 1987, that message received a boost when the team of the racing yacht *Stars and Stripes* regained the prestigious America's Cup for the US, winning in a boat which had a hull covered in riblets.

Another application for riblets appeared in the Strush ST competition swimsuit, marketed by Arena North America of Englewood, California. Arena combined existing riblet technology with innovations developed within the company to devise a swimsuit that has been flume-tested to be 10–15 percent less resistant to drag than any other swimsuit on the market. The Strush design incorporates **silicon ribbing** across the breast and buttocks—the areas most subject to turbulence—which reduces what engineers call hydrostatic resistance. Specifically designed for butterfly, breaststroke, and backstroke, the suits made their debut appearance at the 1995 Pan-American Games in Mar del Plata, Argentina. The US swimmers who wore the Strush suits came away at the close of the Games with **13 gold medals**, three silver, and one bronze.

◖ Designed by aerospace engineers, tiny riblets across the breast and buttocks of this swimsuit reduce drag caused by turbulence. Competitive swimmers enjoy extra speed from the improved flow of water around the contours of their body.

Failures can be just as influential to research as successes—a fact evidenced by the work of pioneering astronaut Eugene Cernan. Although Cernan's attempt to reach a maneuvering unit in space had to be abandoned, it led directly to an industrial application that is today utilized all over the world.

SPACE-WALKING ASTRONAUT IS BLINDED BY CONDENSING SWEAT

NASA SPRAY KEEPS VISION BRIGHT

CLEAR SIGHT

While many controls in modern aircraft are electronically controlled, pilots still need excellent vision, so to prevent their visors from misting up they use a special agent.

FURTHER USES

- Visors for racing drivers and motorcyclists
- Eye protectors for industrial workers
- Car windows in humid climates
- Mirrors in restrooms

In June 1966, astronauts Tom Stafford and Eugene Cernan rocketed into space aboard the Gemini 9 spacecraft. Their mission was to catch and rendezvous with a previously launched target, and to test an **astronaut maneuvering unit** that Cernan would use to propel himself around outside the spacecraft. Two days into the flight, Cernan eased himself into space. He made his way by handrails back across the top of the spacecraft and into the back, where the maneuvering unit was attached. Cernan was to back onto the unit as though it were a chair, strap himself in and release clamps that would allow him and the unit to float free of the restraints.

As Cernan struggled to operate the mechanical linkages before getting on the maneuvering unit, he expended more physical effort than had been allowed for, and began to heat up beyond the capacity of his suit cooler. He sweated inside the suit and his faceplate began to fog, first from the bottom and then toward the top. While he rested, water droplets formed on the inside of his faceplate, and in the hope of clearing them he raised his sun-visor, closed his eyes and turned his head to the full glare of the sun. Slowly the fog began to evaporate, but when he resumed his work it returned. Two hours into his spacewalk Cernan had to **abandon the entire operation**, feel his way back to the hatch and get back inside the spacecraft. The maneuvering unit was never tried in space.

Determined that faceplate-fog would not scupper another mission, NASA improved the capacity of the spacesuit system to handle body heat. Research to develop an **anti-fogging spray** for use on the inside surface of astronaut faceplates was also quickly embarked upon. It was soon discovered that a spray made from a liquid detergent, deionized water and a fire-resistant oil was effective against the build up of fog. In 1980 Tracer Chemical picked up on the technology of the antifogging agent, which had by then become standard kit, and marketed the product under its own name and those of name-brand labels. The company found applications and enthusiastic customers among them fire departments, skiers, and skindivers, and in companies producing car windows, bathroom mirrors, camera lenses, and helmets for motorcyclists and pilots.

COMPUTER TECHNOLOGY AND AUTOMATION

COMPUTERS LIE AT THE HEART OF EVERYTHING WE DO IN THE MODERN WORLD. HOWEVER, WHEN THE SPACE PROGRAM BEGAN IN THE 1950S, THE SCIENCE OF ELECTRONICS WAS IN ITS INFANCY. TO THE BENEFIT OF HUMANS ON THE EARTH, THE SPACE RACE AND THE COMPUTER AGE COINCIDED AND HAVE GIVEN RISE TO MANY TECHNOLOGICAL INNOVATIONS.

Almost everything in space is controlled by computers, either on the ground or on a satellite, but NASA's electronic technology can now be found in a variety of inventions on Earth. From designing comfortable seats (see p. 88) and lifelike robot hands (see p. 90) to virtual-reality holograms (see p. 94) and specially-tailored keyboards for the disabled (see p. 96), NASA computer and automation technology has revolutionized the world in which we live. Even exotic flying machines benefit from space applications (see p. 98), while self-healing computers threaten to outlive their designers (see p. 100).

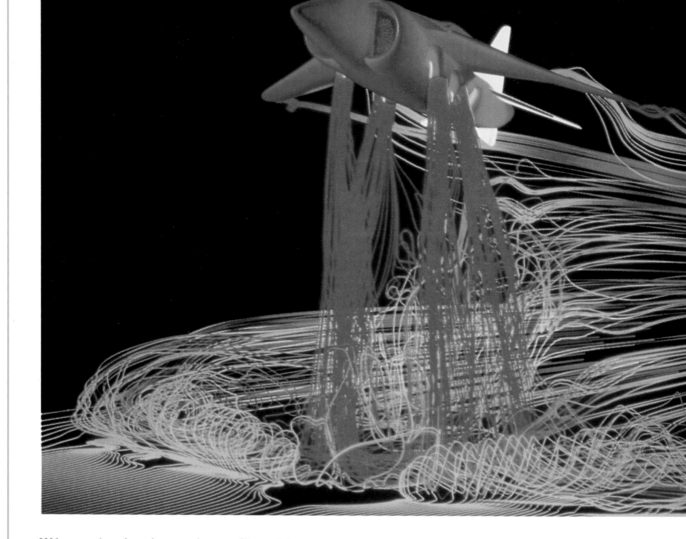

When designing aircraft, ships, or racing cars, today's engineers first create a mathematical model that simulates the way the machine will work. Research by NASA has created computers that can predict how vehicles will disturb the air or water through which they move, and a huge simulator enables information to be assessed quickly.

LINKING DESIGN ENGINEERS

WATCHING THE FLOW

Colored flow patterns mark the turbulent air and vortices produced by this computer model illustrating the way jet exhaust fumes react with air around a vertical jump-jet in flight. By simulating the way different shapes interact with complex flow patterns, engineers are able to refine their designs.

FURTHER USES

- Racing car design
- Touring car shapes
- Streamlines trains and buses
- Faster racing yachts and speedboats

For many years, engineers and scientists have used computers to simulate the way different shapes behave in water and in air. Computers are also used to explore the effect of extreme environments on machines and test their reaction to stresses. To see how an airplane survives in a violent thunderstorm, or how racing cars behave on different tracks, test engineers need tools that can simulate **stresses and extremes** in ways that no wind tunnel can reproduce.

At one of NASA's laboratories near San Francisco, California, engineers have joined forces with mathematicians and computer specialists to assemble the **world's biggest supercomputer**. NASA's supercomputer links 1,400 industry, university, and government users in a high-speed network connecting terminals right across the United States. The machine is the largest computer simulator in existence and is called the Numerical Aerodynamic Simulator (NAS). By using NAS to show how different shapes react to extreme conditions in the real world, a "virtual" picture helps NASA scientists as well as industrial designers improve their designs. In this way, university students, car builders, and boat designers can "test" their concepts in violent weather and high seas without ever leaving their laboratories. To add realism and to display the way wind and water move around different design shapes, the computer network puts color on invisible forces. In addition, to illustrate how the jet exhaust of an airliner disturbs the air behind it, the computer sets it flying through a liquid air, colored to reveal the patterns of turbulence. As the simulated aircraft flies through the simulated air, operators can modify designs to see how changes affect the flow pattern.

In 1993 a new high-speed Cray computer was added to the test network, **raising computing capability** sixfold—but that was only the beginning. As a part of its further development, NASA is now looking to upgrade the Numerical Aerodynamic Simulator to increase its capacity and working speed by over 150 times. When this is achieved, engineers, designers, students, and scientists will have the power to let NAS design its own ships, automobiles, and aircraft, with all its data based on information from the real world.

ERGONOMIC SEAT DESIGN

SPACECRAFT SEATING MOLDED TO THE INDIVIDUAL ASTRONAUT

When seen on a computer screen, Jack appears human enough, but in reality he is a three-dimensional simulator for one of the most remarkably engineered structures of all—the human body. Jack is now hired out to a wide range of different companies and government agencies who wish to redesign their work environments without having to build expensive models and mock-ups.

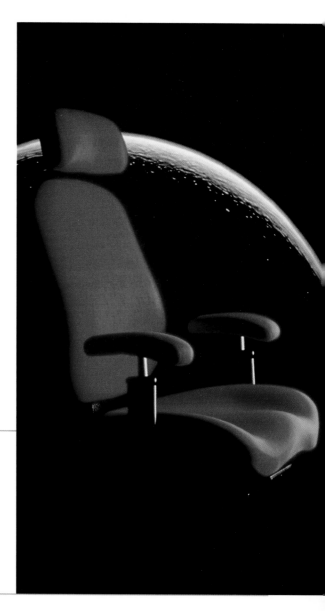

FROM SOFTWARE TO SEAT

The same attention to ergonomic design which is necessary to prevent damage to human body organs during the stresses of a rocket launch, can greatly assist in relieving stress and tension during long periods spent sitting in chairs on Earth.

Jack has the intriguing job of trying out new seat designs for cars, speedboats, and planes, and frequently helps engineers decide how best to arrange controls and instruments. He is completely objective never tires, complains, or takes a lunch break. Jack is, of course, an advanced **human-factors computer software package** designed by the Computer Graphics Research Laboratory of the University of Pennsylvania. In association with NASA, the Laboratory developed Jack to help with a broad range of computer design applications. Now Jack is marketed for use by a wide variety of companies building vehicles of all types. However, it was in space that he was originally intended to find employment.

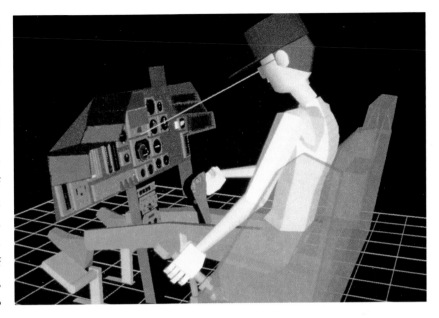

Jack, a computer-generated mannequin, is employed to tirelessly test alternative seat designs for home use, cars, airplanes, and work chairs. A software computer package helps Jack to come up with the most comfortable design.

The pressing need for effective spacecraft seat design came during the late 1950s, when craft such as NASA's one-man Mercury capsule imposed great stress on the human body. In the 1960s NASA began designing advanced spacecraft for operation by several crew members, and the **science of ergonomics** was born. Each astronaut had to be able to reach the controls of the other and cabin layout became a key aspect of spacecraft design.

Jack is a **computerized model of a human figure**, and was designed to put ergonomic design at the forefront of workplace engineering. He consists of 39 segments, 38 joints, and 88 degrees of freedom and has a 17-inch (43-cm) flexible torso. Each separate part of Jack is programed to move automatically in response to commands, and a window on the computer screen shows the view he sees. NASA's Johnson Space Center, which helped to develop Jack, has used him to help work out the best interior design for the modules built as part of the International Space Station. The US Army is putting him to work designing the controls layout for their latest helicopter and the next generation of Army land vehicles. However, in the most promising development of all, Jack has been recruited by another NASA laboratory—the Ames Research Center near San Francisco—to become incorporated into a **revolutionary new project** intended to design machines that will work effectively alongside humans.

FURTHER USES

- High-stress racing seats
- Comfortable chairs for the elderly
- Contour couches for the injured or convalescent
- Support chairs for paraplegics

ROBOT HANDS TAKE THE STRAIN

SURVEYOR USES ITS ROBOT ARM TO DIG UP SOIL ON THE MOON IN 1966

Automated limbs of all sizes have been created for a wide variety of tasks, ranging from handling radioactive fuels and picking through toxic waste, to replicating the opposable human thumb and fingers as prosthetic replacements. The robots now used by industry for these highly complex tasks all employ technology researched at NASA's space centers.

No machine has yet been built with the complexity, efficiency, or discriminating powers of a human being. However, there are many tasks for which humans are not suited, and sometimes getting a human to do them is prohibitively expensive. NASA found that the cost of sending people into space was extremely high, and as a result began sending robots to the planets instead. The first **robot manipulators** in space were carried aboard NASA's Surveyor spacecraft and Russia's Luna series of moon explorers during the 1960s. In the 1970s the US and Russia developed soil scrapers and sample-retrieval arms to recover tiny rocks from Mars and Venus for analysis on board.

During the 1980s robotic arms grew in size and increased their range of functions. The **remote manipulator arm** used for NASA's Space Shuttle is a hefty but precise piece of engineering that has since been used to retrieve and deploy many satellites and platforms, including huge structures such as the Hubble Space Telescope. More recently, a complex set of robot workers went under construction in Canada to help assemble the 400-ton International Space Station.

The Robotics and Automation Corporation (RAC) of Minneapolis, Minnesota has collaborated with NASA's Marshall Space Flight Center to perfect a wide range of machines to perform jobs that humans find repetitive, boring, or dangerous. RAC, who mainly build robots for the manufacturing and engineering industry, has perfected a hand that is **attached to a robot**, and which can be used for grasping tools or pieces of work. RAC also builds work-cells where complete automation processes can be handled in safety. These work-cells contain the machinery and tools for preset tasks whereby the robot changes tools automatically, accomplishing anything from welding, casting, or molding to polishing and finishing. Another important application of robot technology is in the provision of artificial hands that can help disabled people perform everyday tasks.

NASA has provided RAC with valuable information to help it develop **automatic tool-changing** devices using hydraulic, pneumatic, or electric power. One area of manufacturing experiencing a growth in the use of robots is the plastics-forming industry, in which robots are used in the manufacture of products as varied as car body parts and radar domes. Now industry is set to help NASA develop a new generation of robots to be used for repairing satellites in space, without incurring the cost of getting a human mechanic on site.

HELPING HANDS

After four decades of research, robot hands developed by NASA perform highly complex tasks and help disabled people and paraplegics to lead normal lives.

With millions of components being produced every day from the world's manufacturing plants, the use of chlorofluorocarbons (CFCs) in each process has quickly become a major global problem. Testing environmentally friendly alternatives to CFCs using NASA technology ensures that these new products are both green and effective.

DIRT-FREE CHIPS

NASA developments enable silicon chips to be manufactured in a totally dirt-free environment without producing atmospheric pollution.

NASA OBSERVATIONS REVEAL DEPLETION OF THE OZONE LAYER

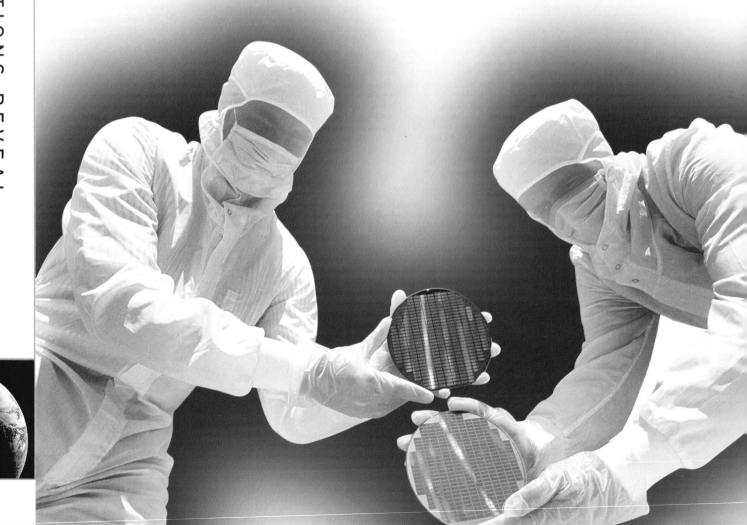

TESTING COMPONENTS FOR THE ENVIRONMENT

A great deal of atmospheric pollution is caused by obvious sources such as car exhaust fumes and fossil-fuel power plants, but the majority of pollutants come from hidden uses in some of the largest manufacturing industries. One of the biggest sources of pollution is chlorofluorocarbons (CFCs), which have been used in almost every industrial process and retail product. In the 1960s NASA discovered that the ozone layer was slowly being **eroded by pollution** from industrial and technological processes and, since then, the space agency has been at the forefront of cleaning up the planet. Nowadays, engineers and inventors regularly call NASA and ask for help in the design of new processes to cut down toxic ozone-eaters.

One such caller was Jeffrey A. Schutt, General Manager of Trace Laboratories-Central, who wanted to know what NASA could do to help him clean up his own particular act. Schutt's company tested electronic components and printed circuit boards after they had been cleaned using an **ozone-friendly alternative** to the CFCs which had traditionally been used for "scrubbing" the components. Trace Laboratories-Central used a different process, but it needed to know that the "green" chemical that had replaced CFCs was not having an adverse effect on the parts.

NASA's work on "electromigration"—the way surface insulation on microscopic components can conduct electricity—helped Trace Laboratories-Central perfect a system of testing that ensured electronic parts and circuit boards left the factory with a high degree of performance and reliability. In this way, Jeffrey Schutt was able to use environmentally friendly products, while maintaining the same quality testing as before. By using NASA technical information, Schutt was also able to further his **research on electromigration** and explore other methods of testing that enhanced the service he could offer to his customers. In turn, this helped to raise the quality control standards in the manufacturing process, cut waste, reduce wasted production time on substandard components, and increase profit margins.

FURTHER USES

o Better manufacturing processes
o Cleaner fuels through removing toxins
o Safer quality-control methods
o Stimulus to biofriendly industries

VIRTUAL REALITY
MIMICS HOLOGRAMS

The first virtual reality and hologram programs developed by NASA were designed to reduce costs in maintaining communications satellites and to ensure the safety of astronauts. In reality the idea was impractical, but the concept and the technology behind it have been developed into some remarkable applications.

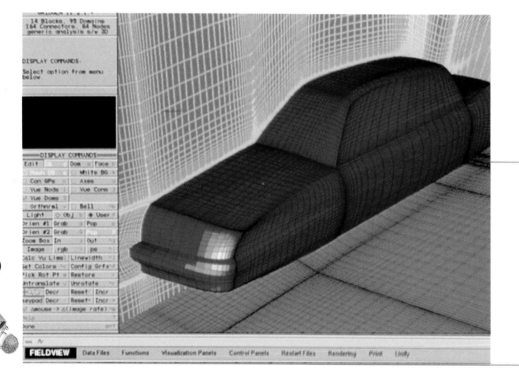

IMPROVING AIRFLOW

Programs developed by Coryphaeus Software enable designers to work with virtual 3D images in order to test wind resistance.

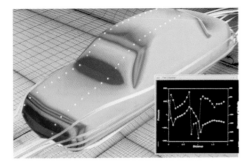

When NASA was in the final stages of building its fleet of reusable Shuttle vehicles in the late 1970s, a variety of applications were planned for the vehicle. Among other uses, it was to have been employed to launch most of the satellites previously sent into space on expendable rockets. Plans were also made for shuttle vehicles to service satellites in space and refuel them with the propellant essential for them to maintain position and **hold the correct attitude in orbit**. Most of the proposed uses proved impractical, but NASA also planned to use the Shuttle for making service calls. It wanted to put astronauts into low-Earth orbit from where they could use a smaller vehicle, like a **space taxi**, to fly the 23,366 miles (36,000 km) to the relay stations that carry most of the world's communications and TV satellites. There, they could carry out repairs and extend the lifetimes of these stations before returning to the Shuttle. However, getting people out that far is costly, and strong radiation is too dangerous for astronauts to spend more than a few hours there.

In the 1980s NASA came up with "telepresence," a **virtual-reality hologram** transmitted from a small unmanned robot to controllers on the ground. From their consoles at NASA's Houston station, engineers would "fly" the robot around the satellite and construct a three-dimensional hologram of it from TV images sent to earth. The hologram would be used to plan a maintenance operation that could be carried out by the robot's manipulator arm working in conjunction with a range of tools and power drives.

In the late 1980s Steve Lakowske founded Coryphaeus Software, Inc., of Los Gatos, California. Lakowske had spent a decade building dynamic motion simulators and human interaction machines at the Ames Research Center, and he applied his experience to high-end, real-time simulation products. One of the most impressive products was a modelling and simulation tool for the development of static and **dynamic 3D databases**. Lakowske moved on to apply his multimedia, multidimensional, virtual-reality world to other software programs and applications: Activation helps companies to conceptualize new computer games; EasyT lets users create large-scale representations of terrain for simulated overflight; while another product helps local government with road layout design.

A major advance from NASA's early holograms can be seen in today's sophisticated programs, which make it possible to produce virtual cross-sections with ease.

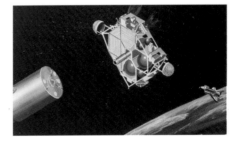

The virtual-reality "telepresence" program was developed by NASA to aid satellite maintenance.

NASA research into improving techniques for programming electronic information in aircraft and spacecraft has led to a keyboard innovation that promises to transform the workplace for hundreds of thousands of disabled people around the world.

ELECTRONIC ADVANCES CONDENSE INSTRUMENTATION ON THE FLIGHT DECK

AIRCRAFT DISPLAY

By using different combinations of keys on this aircraft instrument display, complex information is provided about a wide range of vital systems. The new system saves space, but several keystrokes are required to complete any operation.

KEYBOARD IMPROVES DISABLED JOB ACCESS

The quantity of data the average person has to process is infinitely greater than it was 20 years ago. Almost every job today involves the use of a computer, however distant. Handling, processing, and filing information has moved from shifting pieces of paper in and out of filing cabinets to operating switchboards, keyboards, and on-screen icons. In the case of people who **control complex machines**, for example airline pilots, military pilots, and astronauts, information overload can easily take hold. NASA's search to simplify tasks has its roots in the age of digital electronics during the 1970s, when electrical control systems for aircraft were first introduced. Alternative programing was tested, including **touch-screen loading** and single-key input boards where one- or two-finger operations were all that was required. Computerized information displays soon replaced individual instruments because the sheer quantity of information was too great to display it separately. NASA was also keen that the permanent space stations of the future should contain only computerized displays.

In cooperation with Infogrip, Inc., of Baton Rouge, Louisiana, the Stennis Space Center and the Mississippi State University developed the BAT chord keyboard. Instead of the 101 keys on the standard keyboard, the BAT device contains just seven keys. Five of the keys lie directly under the thumb and fingers of the left hand, and two keys are accessible with a sideways movement of the thumb. Using chordic technology, like that employed to strike a chord on a piano, the operator can cover the full range of options available through a regular keyboard, but with a greatly **simplified finger system**. Operators say that the BAT chord keyboard has ergonomic advantages over the standard keyboard and can be learned as simply as touch-typing on a regular board layout.

In addition to providing a user-friendly system for pilots and astronauts, the BAT chordic keyboard has wider benefits. The single-handed operation of the keyboard is ideal for disabled operators and amputees, enabling **full participation** in all activities involving information systems. It is also particularly suited to those with impaired vision because it is based on Braille.

⬆The BAT chord keyboard allows an operator to use one hand to perform all the functions usually used by two hands on a standard keyboard. The other hand is left free to direct the cursor via a computer mouse.

With increasing demands on their safety and performance, modern airliners bristle with devices adapted from the demanding technology of rocket and satellite design. Today, airplanes can control themselves in flight and even land if necessary, thus removing much of the potential for human error.

DIGITAL FLIGHT CONTROL

The aviation industry is constantly searching for greater maneuverability and agility for aircraft. A significant advance in this field was achieved through the development of automated electronics during NASA's race for space in the 1960s and 1970s. Because most space missions are conducted with robots, NASA had to develop ways for spacecraft and satellites to control themselves **far from human intervention**. The connection with aircraft began when fighter pilots wanted more maneuverability, tighter turns, and lightning reactions from their aircraft. To achieve such a responsive aircraft the design had to be so unstable that flying the plane would be like balancing a lead ball on the head of a pin. The coordinated reactions of the human eye, brain, and hand were insufficient to keep the aircraft from tumbling out of the sky. NASA's Dryden Flight Research Center began tests with **"fly-by-wire"** control systems, which borrowed technology from spacecraft design and married it to sophisticated computers.

"Fly-by-wire" systems did not use direct mechanical linkages; instead the controls in the cockpit were connected by wire to small electric motors attached to the moving surfaces on the wings and tail. As computers became small enough to put into airplanes, the next step was to program an electronic memory to do a job that the average human would find impossible—that is, **balancing the airplane** in flight through thousands of tiny, almost imperceptible, adjustments in attitude every second. Thus was born the "inherently unstable," electronic "fly-by-wire" airplane.

One company at the forefront of this aviation revolution was Smiths Industries, a British company with half its operation in the United States. From its beginnings as a small London clockmaking firm in 1851, Smiths Industries has tracked the evolution of measuring devices and now puts **flight-control computers** into both civil and military aircraft. "Fly-by-wire" technology has allowed designers the freedom to take aircraft performance into realms impossible to obtain through direct manual control. Airliners have now adopted the same technology to relieve pilots of too high a workload.

FURTHER USES

O Digitally controlled racing boats
O Digital sensors for car safety
O Digitally controlled robots
O Computer-controlled unmanned submarines

NEW SHAPES

The shape of the F-117 stealth attack aircraft is designed to make it almost invisible to radar, not for its aerodynamic qualities. In fact, it can only remain in the air through its sophisticated digital flight-control system—a technology that opens the door for unusual aircraft shapes.

COMPUTERS THAT HEAL THEMSELVES

F-22 USES SELF-HEALING COMPUTERS TO REPAIR BATTLE DAMAGE

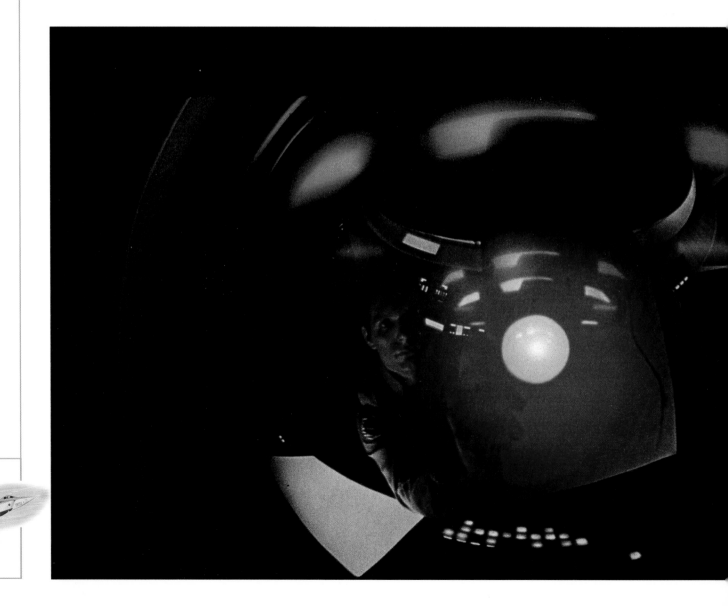

In the early days of space travel, the Moon was believed to be as far as men could travel within a reasonable time. Flights further away would take months and missions to the outer planets would take decades, so NASA developed unmanned probes that could accomplish missions to explore deep space.

The revolution in automated space probes has made flights to distant worlds a reality, but as missions are sent farther away from Earth they require greater autonomy. At the speed of light, it takes just 1.5 seconds to get a radio signal to a robot on the surface of the Moon, but at least five minutes to reach a spacecraft on the surface of Mars and more than one hour to send a message to Saturn. When NASA planned **deep-space missions** to the outer planets they realized it would be no good relying on radio-controlled spacecraft. Robots would have to make independent decisions themselves and not rely on human operators for all the answers.

In a special research program called STAR (Self Test And Repair), NASA engineers attempted to build a system dubbed HAL-2, after the speaking electronic brain that ran the spaceship *Odyssey* in the movie *2001*. STAR was intended to monitor and control the health and safety of spacecraft far from Earth, and resulted in a series of developments that now allow spacecraft to run without the need for human control or intervention. Instead, the new generation of craft are managed by **autonomous computerized "brains."**

The development of **self-repair systems** under the STAR program was soon picked up on. Aircraft builders Lockheed Martin incorporated STAR technology into the super-stealth F-22 air dominance fighter called "Raptor," which is scheduled to enter service in 2003. Raptor has the ability to reassemble its components if knocked out through battle damage in combat, and to repair and maintain its system.

Like HAL in *2001*, intelligent, self-healing machines interact with humans to monitor routine work functions. These computers, however, are far from replacing people in their work.

COMPUTER WITH BRAINS

CONSTRUCTION, TRANSPORTATION, AND MANUFACTURING TECHNOLOGY

NASA HAS CREATED AN ENTIRE INDUSTRY TO DEVELOP A WIDE RANGE OF SATELLITES AND SPACECRAFT, AND MANY OF THE MANUFACTURING TECHNIQUES OF THIS INDUSTRY HAVE BEEN APPLIED DIRECTLY TO THE NEEDS OF PEOPLE ON THE EARTH.

Techniques developed to put men on the Moon have been used to make industry, transportation, and manufacturing more efficient. Basic activities, like cleaning spent rockets, have led to an improved industrial process which has saved both money and time (see p. 106), and the development of ceramic composites for aircraft promises better industrial waste incinerators (see p. 110). Complex solutions to practical problems are borrowed from Apollo management techniques (see p. 116), while quieter and more efficient aircraft engines are promised from a collaboration between NASA and industry (see p. 124).

Nowadays there is little time to wait for discoveries. Instead, engineers must project their ideas beyond the limits of existing knowledge. Computer programs that can predict the aerodynamic performance of a new product before even a prototype has been constructed have helped to streamline the design process.

NASA STUDIES AIRFLOW IN COMPLEX AEROENGINE TURBINES

FLUIDS FOR BETTER DESIGN

⬆ **Aerodynamic flow predictions show how streamlined a skier is when skis are tucked in against the legs and pointed outward, reducing drag and increasing speed.**

THINGS TO COME

Three-dimensional predictions about air flow will help to make future supersonic transportation aircraft more aerodynamically efficient, reducing cost and minimizing noise.

There was a time when it was believed that the key feature of a streamlined object was a pointed nose that would cut its way through the air. It took a long time for engineers to realize that a **streamlined tail** does more than a sharp nose to make an object go fast. A blunt tail causes vortices to sweep in around the back of the vehicle and tug at it like a magnet. By contrast, an elongated tail allows air to flow all around the vehicle without causing turbulence, thus maximizing the efficiency of the engine and making the body more streamlined overall.

One man with a particular interest in predicting the motion of fluids (air and water) in unpredictable situations is Dr. Wayne Smith, formerly of Fluent Incorporated in New Hampshire. Smith contacted NASA's Lewis Research Center in Cleveland, Ohio—an organization that specializes in predicting how flow patterns move in complex turbines and combustion chambers—for advice. In 1987, Dr Smith began work on a computer tool to predict the movement of fluids around both stationary and moving bodies. Dr. Smith wanted to develop a commercial product that could run on a relatively small computer. The result was called Computational Fluid Dynamics (CFD), the fundamentals of which are still entrenched in modern design and engineering. He designed a grid system with tetrahedral (four-sided) cells to divide the flow field into discrete sections, instead of the usual six-sided cells adopted by more traditional CFD programmes. The result is a **three-dimensional graphic prediction** that revealed the way fluids would behave around or through any imaginable structure. In 1992, Smith used his own program, called RAMPANT, to research the way air flows around a skier making a high-speed leap from a ski jump. With the skis set in a V-pattern after the leaving the ramp, the air recombined behind the skier's body just like a streamlined tail.

Dr. Smith has applied RAMPANT to a wide range of applications, including the measurement of air flowing over wings, the way air behaves across flaps and slats to give aircraft extra lift, the flow of gas jets through turbines, and the aerodynamics of vehicles. Now, the Lewis Research Center is using the program to build better engines that burn less fuel and produce **less pollution**.

In just 12 minutes, a waterjet stripping system can strip a segment of Shuttle booster that would take up to six hours by the manual method. Waterjet stripping, which also produces less toxic waste and fewer health hazards for workers, has found a wide variety of applications.

WATERJET COATING STRIPPERS

The two huge, **solid-propellant boosters** that fire for two minutes to lift the Space Shuttle partway into space are returned to Earth by parachute and used again. When they burn out, the Shuttle is about 28.5 miles (46 km) above the Earth, but the forward momentum throws them almost twice that height before they drop back down through the denser layers of the atmosphere. Slowed by parachutes, the boosters splashdown in the Atlantic Ocean, where they are recovered for cleaning and reuse on a later mission.

To protect them from heat produced by friction with the atmosphere as they fall, the boosters have a special protective coating that is **resistant to high temperatures**. However, the coating applied to the boosters is also resistant to stripping, and the manual process necessary to remove it for reapplication was both time-consuming and inefficient.

NASA's Marshall Space Flight Center in Huntsville, Alabama, worked with an industrial contractor to produce a machine that could strip the boosters faster, better, and less expensively. The result was a high-pressure water jet with a **force of up to 6.5 tons per square inch** (1 tonne per sq. cm), the angle and precise force of which must be accurately aligned to strip the coating without damaging

⬆ **When the Shuttle's booster tanks are recovered from the water, a great deal of work still has to be done before they can be reused.**

the metal skin of the booster. At first the robot device was fixed to a gantry at the Kennedy Space Center, but was subsequently fitted to a transportable platform with a plexiglass cabin.

In 1992, Delta Airlines started to use a system developed from a U.S. Air Force device for stripping paint from large transportation aircraft. Called the Automated Robotic Maintenance System, or ARMS, this cleaner cut time and costs by 50 percent. When Delta applied the technology to cleaning engine components after strip-down, it saved them money and **replaced toxic chemicals** like methylene-chloride-based strippers. Further refinements to ARMS make it suitable for cleaning wheels, engine inlets, exhausts, and other inaccessible parts.

In 1995 technicians set up a giant version of the system at the Oklahoma City Air Logistics Center. Exhaustive tests showed that it not only cut stripping time by 50 percent but also **reduced toxic pollution by 90 percent**. As a result, many motor body shops have realized the benefits of the system and are cleaning up their act too.

A CLEAN SWEEP

Developed from a NASA device, industrial strippers replace toxic chemical cleaners in a wide variety of industrial cleaning processes, reducing waste, minimizing the use of harmful products, and saving time and money.

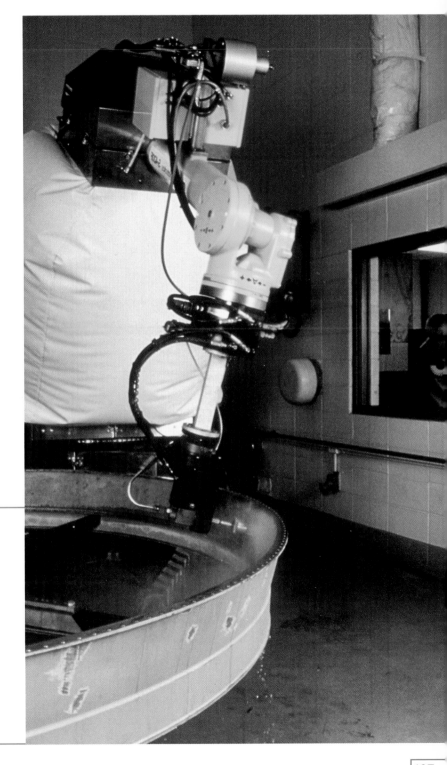

TAGGING ELECTRONIC COMPONENTS

EVERY COMPONENT IN A NASA PROJECT HAS TO BE IDENTIFIED

Personal items that have been marked with a digitally-based identification system are inscribed for life, and their identity can never be erased. From cars to computers, possessions that are invisibly tagged can be returned to their owners if stolen and later recovered by the police.

It takes **several million separate parts** to put the Space Shuttle together, hundreds of thousands of components to build the engines that power it into space, and more than 30,000 heat-protecting insulation tiles to bring it back safely down through the atmosphere. Someone has to keep track of all those parts because if anything goes wrong, even a seemingly insignificant error in an electronic resistor, inspectors will want to know the detailed history of that component. Moreover, each tile is cut uniquely for its specific location on the Shuttle orbiter, and should it need replacing its exact size, shape, and thickness must be known.

For years NASA wrestled with the problem of parts tagging. **Bar-code systems** required too much surface area on which to install the label, and any tagging system chosen had to withstand a temperature range from -284°F (-140°C) in the blackness of space to more than 2,192°F (1,200°C) on reentry. During the 1980s NASA's Marshall Space Flight Center examined the solutions that were being applied in other industries to meet similar needs. The result was a **microetching system** that has subsequently been used in the pharmaceutical, electronic, and automotive industries. NASA applied the high-data, high-density, machine-readable tagging system to Shuttle component inventories, and used digital data matrix

⬆ **This two-dimensional bar-code tag can be miniaturized and then applied by microetching. It bears a full record of the component's manufacturing history and provides a trace in the event of failure.**

technology to store and record full histories of each part. Through a special agreement with NASA, in 1997 the CiMatrix Corporation of Canton, Massachusetts, set up a special Symbology Research Center (SRC) at Huntsville, Alabama, near the Marshall Space Flight Center. SRC produced a tag system with invisible and virtually **indestructible identification** etched by laser on areas as small as .00015 of an inch (.0004 of a millimeter).

The SRC tagging system is a cost-effective way of keeping records, tracking each part in complex machinery, and maintaining **complete performance charts**. The system works on steel, metal, plastic, glass, paper, fabric, ceramics, and composites, and can be applied to any product regardless of shape or size. SRC tags can survive wind, rain, sleet, snow, ice, and heat, and will never peel or come off.

KEEPING IT SAFE

A wide range of domestic, commercial, and industrial products can be tagged with their owner's reference as a trace in the event of theft. Tagging deters thieves and reduces insurance costs.

COMPOSITE CERAMICS IMPROVE EFFICIENCY

NASA's work on ceramic composite coatings has led to the manufacture of a wide range of new materials that can maintain their rigidity and performance in both supercold and superhot temperatures. This technology has found a wide range of industrial uses, from tiny integrated circuits to jet engine turbine blades.

In the early 1970s aircraft builders began to make use of new light but strong materials called **carbon-fiber composites**. This material was first introduced on military aircraft like the Grumman F-14 Tomcat, a plane that is still a front-line air superiority fighter with the United States Navy. Carbon-fiber was then developed by NASA to build tough structures where light weight was a prerequisite for successful operation. Yet for all their advantages, carbon-fiber composites are brittle and have a tendency to propagate fractures and cuts.

NASA's Lewis Research Center wanted to find suitable composites for high-temperature parts in advanced jet engines of the type supersonic airliners would later use. As part of a small-businesses innovation program started by the Lewis Research Center, the Advanced Ceramics Corporation (ACC) of Cleveland, Ohio came up with a new coating for ceramic composites that makes them **one thousand times more durable**. Ceramic composites are made by reinforcing refractories with **high-strength ceramic fibers**. Refractory materials do not distort or melt under very high temperatures and are commonly found in turbine blades or engine compressors that have to survive enormous stress and extreme heat. Composite materials with refractory properties are a great deal more efficient, not

LONG-LASTING GUINEA PIG

The agile X-29 was one of the first aircraft to use revolutionary carbon-fiber composite materials, and forward-sweeping wings for increased efficiency.

only in aircraft engines but in gas-fired power turbines, industrial boilers, waste incinerators, and other high-temperature applications. Aircraft engines with composite liners and shrouds are much lighter and permit greater payloads or increased range, while industrial turbines lined with this material produce **greater heat levels** for the same amount of energy. Composite linings have also been shown to save fuel and increase output. NASA and the aircraft industry are particularly interested in the application of this technology because of the increasing demand for bigger aircraft with longer range or greater speed, producing shorter travel times.

To give the ceramic composite a resistance to temperature fluctuations and moisture erosion, a **special protective coating**, which allows the material to be used at higher temperatures, is applied. This coating, a boron nitride deposit, greatly enhances the effectiveness of the ceramic and gives it properties that provide added benefits. For example, one unexpected bonus is that the coated composite turns out to be an excellent electrical insulator.

FURTHER USES

- Increased super-conductivity for electrical components
- Turbine blades for jet engines in civil and military aircraft
- Greater heat resistance for industrial waste incinerators

111

How many times can you bend a paperclip before it breaks? Even everyday objects require testing for manufacturers to produce a safe, reliable product. Who better to develop ways to test materials than NASA, whose everyday products are high-tech machines?

TESTING MATERIALS FOR SAFETY

Testing the strength of materials is just the beginning of the story. We need to know a great deal more about manufacturing components—in particular, we need to know their age before they reach the production line. Things wear out and break, so to **prevent aircraft from falling to pieces in flight**, or ships' hulls from coming apart at sea, it helps to know, before they are put into service, just how long they can be expected to last. Wearing out of materials is a fact of life, and engineers and technicians involved in manufacturing, assembly, or construction spend much of their time analyzing the effects of stresses and strains.

Things wear out in space too, although there the reasons why they erode can be a little different. For instance, the environment is different, temperature fluctuations are enormous, and **radiation can cause havoc with materials**, bringing about premature decay. To research and understand these problems George E. Caledonia and Robert H. Kresch founded Physical Sciences, Inc., (PSI) of Andover, Massachusetts. The company was set up to help engineers and technicians in the satellite and space manufacturing industry gain detailed information

A paperclip is an everyday object we all rely on. An engineer has calculated the stresses it endures, to provide a product that we can depend upon.

EVERY NASA CRAFT IS STRENUOUSLY TESTED BEFORE IT GOES INTO SPACE

about life-limits on materials. Through a device called FAST—the Fast Atom Sample Tester—PSI was able to **simulate conditions in the vacuum of space** and measure the corrosive effect of atomic oxygen in orbit. PSI performed accelerated erosion and decay testing to show the effects of long-term exposure on a wide range of materials, thus providing satellite manufacturers with information they had needed for some time. Quite soon, however, PSI's task became more than just the initial single application for which it was researched and developed. The company is now applying its work to some very down-to-earth problems.

In 1992 PSI developed the Optical Temperature Monitor, or OTM, a device which is used to measure the temperature of materials known to have emission qualities that change with time. OTM is capable of **accurately determining to a fraction of a degree** temperatures of materials in the range of 1,290–4,530°F (700–2,500°C), and has found application as a device for measuring and controlling the temperature of gases in industrial boilers and home heaters. The basic thermostat in any home boiler is, therefore, more than likely to incorporate a sensor that was originally developed from a very different space product.

CONTROLLING APPLIANCES

The Optical Temperature Monitor (OTM) conducts a test to detect minute temperature differences in a wide range of domestic and industrial processes. OTM can help improve efficiency and can also save energy.

Risk assessment is what the space program is all about. Trade-offs between performance and the possibility of failure are a daily necessity for program managers and astronauts. NASA's research into risk assessment, while never claiming to prevent human error, has made a significant contribution toward making the world safer for us all.

MINIMIZING DANGERS

The collapse of part of Connecticut's Mianus Bridge in June 1983 showed that, even with high safety standards, a more stringent assessment of risk is essential to reduce failure rates.

WITHOUT RISK ASSESSMENT, NO NASA MISSION COULD EVER LEAVE THE PAD

RISK-ASSESSMENT TOOLS SAVE LIVES

We live in a dangerous world. Every day the news reminds us of the harsh consequences of mechanical failure, human error, and risk. It may seem that we have little or no control over events, but the Lockheed Engineering and Sciences Company disagrees. They claim that the solution lies in comprehensive risk assessment followed by adequate **risk management**. And if performance is anything to go by, Lockheed may be right.

Space rockets involve risks that are difficult to assess: these firecrackers work only once, and there is no successful way of duplicating all the different **stresses and strains** they encounter on their way into orbit. These risks require a complex and sophisticated analysis based on sound mathematical formulas and precise understanding of the problems involved. Lockheed produced a sophisticated software system for evaluating risks, and a design team led by James Miller developed the Failure Environment Analysis Tool (FEAT). Using what are known as directed graphs, or diagraphs, to show cause and effect in a string of events, FEAT was developed with help from NASA and applied to broader uses. The system highlights failures using charts and diagrams that are familiar to the operator, and it shows how failures propagate through a system. FEAT can, in essence, accept any program or system and assess its risk. By **using real failures** to demonstrate the system, its methods can be applied to predict uncertainties that could cause failure in seemingly safe systems such as industrial heating units, aircraft refueling operations, or bridges.

In 1991 James Miller left Lockheed and set up a company called DiGraphics. Miller, together with Mike Austen (a senior computer programer from the Lockheed-NASA team) and marketing executive John Reed, devised the Diquest Analyzer in 1992. Initially aimed at risk assessment in the chemical industry, the product has found applications in monitoring complex operations that involve several separate but parallel functions with failure threats. It has also been used for **diagnosing fault paths** and training new personnel. Diquest Analyzer highlights where redundancies have been built into a system and tracks where failures can occur in spite of precautions.

PROJECT MANAGEMENT PLAYED A VITAL ROLE IN APOLLO'S SUCCESS

The Apollo mission put men on the Moon in under eight years, and the efficiency with which that task was conducted has been applied to many other projects. One of the most notable applications of Apollo technology has been in highway planning and the management of traffic along major stretches of urban freeway.

APOLLO TECHNOLOGY
KEEPS TRAFFIC MOVING

In 1985 Federico Pena, the US Secretary of Transportation, formally opened the control center of the Transportation Guidance System (TransGuide) at San Antonio, Texas. The first, and still the most advanced, **intelligent highway system** in the world, TransGuide uses management concepts pioneered by NASA to reduce congestion, eliminate the need for wider roads, and cut down on pollution. It works by integrating sensors, message signs, cameras, and computers into one coordinated system that manipulates the flow of traffic to reduce bottlenecks and keep vehicles moving.

The TransGuide project began in 1990, when a major traffic study of the San Antonio freeway system suggested more intelligent ways to use the existing road system without building additional traffic lanes in an expensive widening program. After a competition among several **major high-tech firms**, the Department of Transportation gave Allied Signal, a major NASA contractor, the job of completing the design, buying and installing equipment, and training personnel to manage traffic on what would eventually be a 190-mile (300-km) freeway network. It also involved managing more than 60 companies in specialized areas.

The TransGuide system includes pairs of vehicle detectors in each lane at intervals of about 2,624 feet (800 m), full color video cameras at intervals of 1 mile (1.6 km), 50 fiber-optic message signs, 358 overhead lane control signals, and a computerized traffic light control system. At the heart of the system is a 48,000-square-foot (4,460-sq-m) **management center** where sensor signals, video images, and message board indicators are integrated and displayed on a video wall almost 65 feet (20 m) square. Each operator has four video consoles, and monitors the automatic adjustment of traffic-flow conditions to maximize use of the road system. Operators can override the system and distribute the traffic according to specific requirements.

SOLVING PROBLEMS

TransGuide detects, locates, and categorizes a traffic incident within two minutes and works out a rerouting solution within 15 seconds. It will also automatically inform drivers of detours.

LAMINAR-FLOW WINGS IMPROVE PLANES

Air moving over a conventional aircraft wing tends to become more turbulent the faster the plane moves. This causes "induced drag," a tendency for the turbulent air to stick to the wing and slow it down, and reduces lift. When worked on developing better and more efficient wings, they developed a new type of light aircraft.

During the 1970s NASA's Langley Research Center conducted studies on improved wing profiles. The aim was to develop a wing design that would provide greater lift from a surface shaped to avoid the flow of air breaking up into swirls and turbulence as it passes from the leading edge to the trailing edge. The designs put to the test were known as laminar-flow, and provided the key to achieving better performance for combat and high-speed reconnaissance planes. Laminar-flow wings were tested at NASA's Dryden Flight Research Center during the late 1970s, and by the early 1980s research showed a 30 percent **reduction in induced drag** and greatly increased lift.

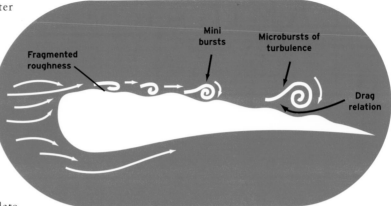

This diagram, which has been exaggerated for effect, shows how air can break up into swirls as it crosses rough and uneven surfaces along the top of the wing, thereby creating vortices and drag that greatly reduce efficiency.

All this was good news for Tom Prescott and Stan Blankenship of Wichita, Kansas, who were designing an affordable, **high-performance** light aircraft. Dan Somers from NASA Langley designed a wing for the new airplane, incorporating the latest laminar-flow research to provide greatly enhanced performance and handing qualities. Prescott Aeronautical Corporation was formed in 1983, with the specific purpose of producing kit parts for the new design. By incorporating an engine in the rear of the fuselage driving a propeller at the back, the corporation created the Prescott Pusher, which was tested at Wichita State University.

The little Pusher was the first airplane to use **computer-aided design** and manufacturing techniques for the hard tooling that fabricates the kit parts. The Pusher made its first flight in July 1985 and appeared at the Oshkosh air show later that year. It was followed by a second prototype in 1987. The first kit-built Prescott Pusher flew in 1988 and proved capable of reaching speeds of almost 200 miles (320 km) per hour. Many aircraft now have laminar-flow wings, producing enhanced performance, greater efficiency, reduced fuel use, and increased safety.

EFFICIENT AND ECONOMICAL

Laminar-flow light aircraft made from composite materials are much more efficient in the air, needing smaller and less powerful engines to achieve the same performance, thus lowering costs for the operator.

FURTHER USES

- Laminar-flow racing boats
- Laminar-flow car bodies
- Improved flow in smooth turbine engines
- Better streamlining on trucks

"NEWTSUIT" HELPS DIVERS

Exploring for natural resources is a vital part of supplying society with electricity and power. However, the effort of retrieving vital fuels from below the oceans has taken technology to the edge. Technology developed in space has been put to use underwater: Undersea divers are now better protected because of developments in Moonsuit technology.

Divers need to spend many hours working at great depths to secure oil platforms. Similarly, the machinery of the gas industry runs only because of the efforts of an undersea workforce, operating in extremely **dangerous conditions**. Many of the tasks involved in underwater inspection and the maintenance of offshore oil rigs are complex and cannot be undertaken by automated systems. So it was with great interest that Phil Nuytten, President of Can-Dive Services Ltd, based in North Vancouver, British Columbia, Canada, came upon studies by NASA's Ames Research Center at Moffett Field, California, into hard-suit technology for Moon workers.

In the 1960s the Ames Center developed a rigid spacesuit that would protect the members of a lunar-based workforce from micrometeoroid impact. The suit would also be able to provide a **pressurized environment of oxygen and nitrogen** (air), duplicating the atmosphere of Earth at sea-level pressure. It was hoped that this suit would enable Moon workers to put in a day's toil without developing serious health problems. However, the Ames hard-shell pressure suit looked more like a personalized spacecraft than a conventional spacesuit. It comprised a rigid torso, limbs, and helmet, and incorporated fluidic joints that allowed a **freedom of movement** denied to astronauts wearing

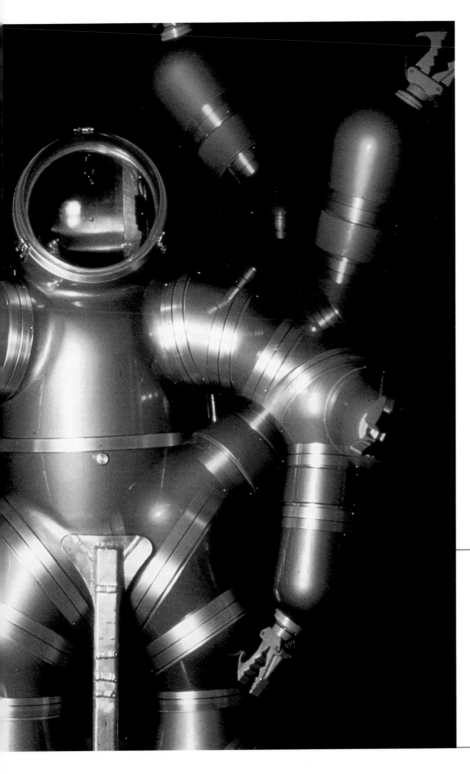

flexible suits. The Ames suit's combination of rigidity and air-breathable atmosphere appealed to Phil Nuytten. He would later design a similar suit, called the "Newtsuit," which allowed divers to work at depths of 985 feet (300 m) breathing air at sea-level pressure.

The Newtsuit is made of aluminum, weighs little more than 395 pounds (180 kg), and incorporates a backpack breathing system with a 48-hour supply of air. With a series of **patented low-friction joints**, the Newtsuit makes underwater movement easy, and the ability to breathe air at sea-level pressure has enormous advantages. Usually, a dive to 655 feet (200 m) under ambient pressure requires the diver to spend five days or more in a decompression chamber to avoid decompression sickness, known as "the bends." This process is financially beneficial for the oil operators, too, because divers are not paid solely for their worktime but also for the total time they spend in decompression chambers.

DEEPER AND DEEPER

Deep-sea divers are limited by the ability of their lungs to withstand the pressure of the water, but hard suits capable of providing a reduced pressure internally can allow divers greater access to deeper regions.

It is not unusual for a good idea to get developed, but fail to make it into production—or at least not for its original purpose. Whenever you boot up a computer, you have a failed NASA technology to thank for a system that never made it into space but found a useful home on Earth.

MAGNETIC LIQUIDS HELP COMPUTERS

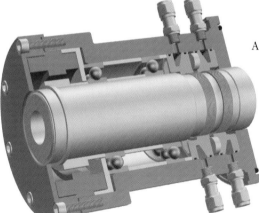

🔼 **This cutaway diagram shows (in yellow) a leakproof ferrofluid seal. Magnets (red) produce a magnetic field that pulls the liquid into contact with the rotating shaft.**

A magnetic liquid, or **ferrofluid,** was developed by NASA's Lewis Research Center in Cleveland, Ohio, to find a way in which weightless propellants could be channeled into the engines of an orbiting spacecraft. On the Earth, fuel settles at the bottom of a tank because gravity pulls it toward the Earth. In weightless space, however, liquids drift around in globules and will not fill the drain pipes that lead to the engines. Something is needed to pull the fluids down into the bottom of the tanks. The logical solution was to add **tiny particles of iron oxide** so that the fuel could be pulled toward a magnetic source situated at the entrance to the fuel pipes. As soon as the engine started it would create the effect of having its own artificial gravity and, in so doing, would provide the acceleration to push the propellants down into the delivery tubes.

NASA was never really sold on the idea of a magnetic system, and relied instead on tiny thrusters to provide a positive acceleration just before the main engine fired. Two NASA scientists, Dr. Ronald Moskowitz and Dr. Ronald Rosensweig, saw potential in another application, and used ferrofluids to create a leakproof seal for the **rotating shaft** of a system used in making semiconductor "chips."

The ferrofluid solved a persistent problem of contamination due to leaking seals, and the two scientists left NASA in 1969 and set up the Ferrofluildics Corporation of Nashua, New Hampshire. The company grew as the market expanded, and a subsidiary, SPIN Technology, was formed to produce a **spindle for rotating computer disks**. The ferrofluid spindle, which uses a film of magnetic fluid instead of conventional ball-bearings, offers greatly increased rotational stability, meaning substantially reduced vibration and mechanical noise. The system also enables disks to store up to 10 times their previous volume of information. The majority of computer=memory disk drives now employ **magnetic fluid-exclusion seals,** and ferrofluids are applied to a wide range of devices, including robotic systems, fiber-optic assemblies, and laser systems.

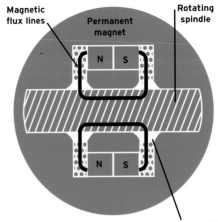

Magnetic flux lines
Rotating spindle
Permanent magnet
N S
N S
Magnetic fluid

Shown diagramatically, the principle of the ferrofluid seal involves a liquid that follows the flux lines of a magnetic field and is pulled into close contact with a rotating spindle.

INDUSTRIAL APPLICATIONS

In Britain, Liquid Research Ltd. and Advanced Fluid Systems have used ferrofluids in a range of motorized shafts designed to operate in a vacuum chamber or an ultraclean environment. A magnetic fluid seal around the shaft prevents air and particles from entering the chamber.

CLEANER AND QUIETER AIRLINER ENGINES

NASA TESTS ENGINES FOR NOISE AND EFFICIENCY

Thanks to decades of collaboration between NASA and the airline industry, the latest generation of airliners produces less pollution around airports than the cars that bring the passengers. Airliners now contribute less pollution to the world's atmosphere than any other form of transportation, industry, or energy-production source.

NEW GENERATION

A British Airways plane flying the CFM56 engine—cleaner, more efficient, and with far less toxic emission than standard turbofan engines.

Much of the technology developed for rocket engines in the space program has fed across to the design and development of aircraft engines. In the late 1960s NASA's Lewis Research Center, which was responsible for propulsion concepts and engine design, initiated the Quiet Clean Shorthaul Experimental Engine (QCSEE) program. The program, which focused on technologies to **lower engine noise and reduce toxic contaminants**, carried on into the 1970s and was largely successful. It resulted in noise reductions of 8–12 decibels, the equivalent of 60–75 percent, for the type of engine used by a Boeing 747. As a result, the US aeroengine maker General Electric conducted a joint programme with NASA for an Energy Efficient Engine (EEE) that was designed to reduce fuel consumption, reduce emissions, and lower maintenance costs.

General Electric began work on an international project, the GE90 turbofan engine, which made its debut in 1995 aboard a Boeing 777 airliner operated by British Airways. The GE90 was developed in cooperation with Italy and Japan, and was not only one of the most powerful engines ever built but also one of the **most efficient high-thrust jetliner engines**, setting new levels of economy and environmental acceptability. Prior to the launch of the GE90, General Electric had taken the original research work it had conducted with NASA into cooperative international engine programs such as the CF6 and the CFM56, which power hundreds of aircraft around the world. This in its turn stimulated awareness of environmentally friendly, clean, and quiet engines, both within the industry and among the general public. It also led to Airbus Industries adopting engines that provide significant improvements on those that were in routine use until the early 1990s.

Technically, the GE90 series has an **inherent design approach** that lowers noise and improves efficiency. Modern turbofans take in air that is compressed, burned in a combustion chamber, and expelled to generate power for driving the turbofan and the compressor. A much greater quantity of air bypasses the combustion process and is pushed back to mix with the hot air coming from the exhaust, increasing thrust and reducing fuel expenditure.

The "bypass ratio" of the GE90 is one way of improving engine efficiency, and the engine's **huge multibladed fan**—with a diameter of almost 10 feet 6 inches (3.2 m) and a bypass ratio of 9:1—further reduces noise and fuel consumption. In specific terms, the GE90 has 10 percent lower fuel consumption than earlier airliner engines, and has been tested to reduce emissions by between 25 and 60 percent.

⤊ Before being installed in an operational airliner, a GE90 engine is put through its paces on a test rig.

INDEX

ACKNOWLEDGMENTS

Picture credits

r = right, *c* = center

Advanced Fluid Systems/Liquids Research Limited 122*c*, 123; Art Directors and Trip Photo Library 24*r*, 25; Breitling SA 32*r*, 33; British Standards Institution 113; Austin J. Brown 118*c*, 125; Paul Finch 70*r*; Ronald Grant Archive 100*c*, 101; Greenpeace 42*c*, 51; Morgan Kaolian/Aeropix 114-115; Outside In 29; London Fire Brigade 30*c*; Natural History Photo Agency 56*c*; Positive Systems 41; Reebok 76*c*; Murray Schwarz 62*c*, 63; Science Photo Library 36*r*, 37, 69, 90-91, 92*c*; Sensormatic 68*c*, *r*; Smiths Industries 96*c*; Still Pictures 79; Telegraph Colour Library 72*c*, 73; Tempur–Pedic (UK) Ltd. 14*c*, 15; Waterjet Systems, Inc. 105; all other images provided by NASA.